Baby Space

ELLEN LIMAN

Baby Space

FINDING, FURNISHING, DECORATING, AND EQUIPPING A PLACE FOR YOUR NEWBORN

Illustrations by Sheila Camera

Quill

A HarperResource Book
An Imprint of HarperCollins *Publishers*

The author has used her best efforts to include information that is as up-to-the-minute as possible as of the date of printing. However, readers are encouraged to consult product guides and independent product reviews to ensure that readers' choices of furnishings and accessories meet all applicable safety standards.

Excerpt from "The Girl That I Marry" by Irving Berlin reprinted page 97: © copyright 1946 Irving Berlin; © copyright renewed 1974 Irving Berlin. Reprinted by permission of Irving Berlin Music Corporation.

The author gratefully acknowledges permission to quote from the following sources:

McGraw-Hill Book Company for an excerpt from *Growing Up Free* by Letty Cottin Pogrebin, copyright © 1980 by Letty Cottin Pogrebin, published by McGraw-Hill Book Company.

David McKay Co, Inc. for an excerpt from *Now That You Have Had Your Baby* by Dr. Gideon Panter and Shirley Motter Linde, copyright © 1977 by Dr. Gideon Panter and Shirley Motter Linde, published by David McKay Co, Inc.

Pomeranz, Virginia E. and Dodi Schultz for excerpts from *The First Five Years: A Relaxed Approach to Child Care* by Virginia E. Pomeranz, M.D., with Dodi Schultz, copyright © 1973 by Virginia E. Pomeranz and Dodi Schultz.

This book was originally published in 1983 by Perigee Books, a division of The Putnam Publishing Group. It is reprinted here by arrangement with Penguin Putnam Inc.

HarperCollins books may be purchased for educational, business, or sales promotional use. For information please write: Special Markets Department, HarperCollins Publishers Inc., 10 East 53rd Street, New York, NY 10022.

First HarperResource edition published 2000.

Library of Congress Cataloging-in-Publication Data has been applied for.

ISBN 0-06-095627-5

00 01 02 03 04 QK 10 9 8 7 6 5 4 3 2 1

To Arthur of blessed memory and our babies Lewis, Emily, and Douglas, and to Abagail and the next generation of Liman babies— all the inspiration for this book.

Contents

Part Three: Fixing Up Baby Space

Part Four: Baby Space Planner

Introduction

The great day nursery, best of all
With pictures pasted on the wall
And leaves upon the blind—
A pleasant room wherein to wake
And hear, the leafy garden shake
and rustle in the wind . . .

ROBERT LOUIS STEVENSON

The joy of discovering that you are having a baby is often followed by panic—"Where will we put the baby?" The timeless problem of housing an infant has become more difficult to solve in an inflationary era. The days of large homes with extra rooms that could accommodate many children are forever lost.

Finding the right place for a nursery and then fixing it up is a challenge even to those with money and space to spare. After all, this will be the busiest room in the house and the center of your child's many activities—at first, sleeping, eating, bathing, playing, and then studying and entertaining—so its location and design must be carefully planned. In addition, as all new parents discover, providing space is more than finding sleeping room. Babies come with extraordinary amounts of equipment and playthings. As quickly as they take over all your attention, their possessions take over the house.

This book is intended to take some of the fear and financial worries out of the arrival of a child by suggesting ways to create economical, functional, and flexible space for a baby. It is not necessary to move or add on to a house in order to have room for a baby. With more creativity than money, you can have a space that is attractive, expandable (for siblings, friends), and adaptable—able to change as quickly as a baby does (space for diapers, then toys, then books). With planning and know-how your baby can be brought up in an environment that is stimulating, safe, easy to maintain, and designed both for your convenience and your baby's comfort.

The principles of decorating a child's room are different from those applicable to the rest of the house—which is the reason for this book. Furniture for adults can last indefinitely, but nothing becomes obsolete faster than a crib, a play yard (playpen), or nursery wallpaper. Fortunately, there are techniques for avoiding the expense of constantly redecorating by planning at the outset for space that will grow with the child rather than be outgrown by him.

Finally, this is an idea book, not a manual of detailed instructions. Operating on the assumption that the parents of today are too busy with other responsibilities, I have described quick decorating tricks that can produce maximum effect with minimum effort and ability. In conveying—through text and illustrations—the experience of hundreds of parents who have overcome space constraints in innovative ways, my intent is to stimulate the reader to find his own solutions. For no matter what your space or budget restrictions are, there is room for your baby.

Finding Baby Space

Making Room for Baby: Some General Guidelines

Selecting the Space

Babies are more resilient in their living requirements than we realize. A tiny baby who has just come from the warmth and security of a small space will need only a small space in the first few months—a place that is safe, warm, clean, light, and airy, to sleep, be fed, changed, and bathed. And although it should be somewhat segregated, it need not be totally quiet. Many babies are abruptly born into bright lights and live their first few days in a frantic, far-from-quiet hospital nursery that is good training for living under less than perfect conditions. According to some pediatricians, children learn to live with such household noises as ringing telephones and only react to change—if the ringing stops, for example. In fact, general household sounds such as the hum of a vacuum cleaner or air conditioner can, it has been shown, have a soothing rather than a disturbing effect.

Ideally, the nursery space should be in fairly close proximity to parents and to the kitchen or bathroom. In the absence of a separate room, and according to the above requirements, the nursery can be set up in almost any vacant spot, including the corner of a living room, dining room, or master bedroom. Some parents find sharing their room—and sometimes their bed—with the baby a more intimate experience, and a great convenience if the mother is breast-feeding. Others become too anxiously attentive to the baby and find that such constant close proximity to the baby interrupts what little sleep they are able to get.

Planning Ahead

Designing an environment that will be constantly evolving, for a client whose gender and personality may not yet be known, is a true test of ingenuity. On the one hand, there's an advantage to furnishing a short-term semipermanent space. Handy parents can experiment without feeling inhibited, without spending too much money, and without comments or critiques from their child (this will come soon enough). On the other hand, if you're not handy and the same space may have to last through all your child's growing years, certain furnishings chosen now for the newborn should suit the future toddler and teenager. At the moment, this may be hard to appreciate. But children and their needs do change "overnight." That now immobile bundle in the bassinet will soon require running-around room, safe play and work space, less storage space for large baby equipment, more for toys and games.

In many cases, it is wiser to invest in the best-quality furnishings you can afford rather than small-scale, strictly baby furniture. A sturdy carriage or crib will hold up through subsequent babies and a chest of drawers of ageless design will hold a lot and last a lifetime. However, I would not recommend buying too much furniture at first or until you understand your storage needs. A chest may not even be necessary; a few shelves in the closet may provide ample storage for a small baby's clothes. For tight budgets it may make better economic sense to buy less than the best furniture and equipment, or to borrow it. For example, cardboard chests are fine for a temporary nursery and can be discarded without guilt feelings when you move, need more storage, or have more money for furniture. Much of the moderately priced, strictly baby furnishings or equipment available today are quite adequate and safe, for the short time they are in use.

Instead of planning a room with the older child in mind, you may find that you are doing the opposite—adapting an adult room for your baby, as one smart woman did with her den. "I started with a grown-up room that had two armchairs, a sofa, and breakfront," she says. "I added a crib, baby accessories, put a changing pad on the desk portion of the breakfront (the rest of it holds baby things), replaced the sofa with a sleep sofa and recovered the chairs with cheerful, washable slipcovers. When I need more space, I will remove the armchairs and when the baby grows out of the crib, it can sleep on the sofa, which is really a slip-covered high-riser bed."

Major decorating decisions involving backgrounds—floor, walls, and ceilings—can be made before the baby is born, even if its gender is not known, although today through amniocentesis it often is. Since these areas are more nearly permanent and costly, you may want to make them neutral and thus adaptable for a child of another sex, age, or, for that matter, an adult. In that case, add furnishings that can be short-term, selected expressly for this baby, such as nursery coordinates, curtains, and throw rugs. Later, as the function of the room and the age of the child change, they can be easily removed.

If the present backgrounds are in good condition, it's practical to change only the fabric or accessories of the room now, waiting until the backgrounds need renewing or the baby begins to show a personality that suggests a decorating theme. Parents who have gone overboard at first have often found themselves redoing the room within the first year or two in a more suitable style. "The mistakes I made, I made before 'I met my child'—I should have had a hard and washable floor covering because she is very messy and an avid artist who likes to work on the floor. I should have installed more shelves for the storybooks she loves to read."

Since babies have been known to arrive early and new parents have been known to have little time for decorating projects, try to do major work ahead of time, if you are not

multipurpose rooms

den at night, children's playroom by day:
toy and equipment storage behind curtain,
Formica-topped coffee table with child-size chairs,
seating banquette, and storage for playpen and stroller

superstitious. Although many a new father has dashed home from the hospital on delivery day to paint or paper, it is better for the baby—no fumes or mess—and for the father, to get this out of the way early. Baby furnishings that personalize the room and add color, such as pictures, curtains, bedspreads, chair pads, bedding, and wallpaper, as well as coordinated nursery furnishings (crib skirts, canopies, pillows, bedsheets, diaper stackers, lamps) can all be ordered in advance and delivered, if you wish, after the baby is born. Just double-check with the store to be sure that essentials, such as crib and sheets, arrive at home before you and the baby do.

The last word on planning should be had by children. Since you will be laying the groundwork now for the day in the not too distant future when the room you have created will be judged by your child, I thought you might like to know what-kinds of comments you can look forward to:

"I would like a very airy white room with carpeting, blue checked curtains, bedspreads, and pillows, a white Formica desk and a blue easy chair."

"I love the green shutters on my windows. I do not like all my brother's toys on the floor."

"I like my room because it's big with lots of shelves. I don't like my room because my shelves are a deep yellow. I want them light purple."

"I don't like not being able to lock my door."

". . . and I want a niche in the wall for my bed with a ladder."

"You have a bunk bed. You play on the top. You sleep on the bottom. And a slide goes from the top to the floor."

"I think there should be three white walls and one blue wall. The blue wall should have windows and plants. I like bunk beds and plenty of storage space. There should be room for posters."

"My ideal room has a white canopy bed, a white shag rug and modern furniture and flowered multicolored wallpaper."

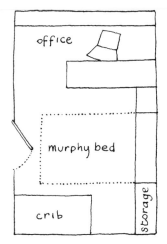

at home
Office.
nursery.
guestroom.

rooms that grow

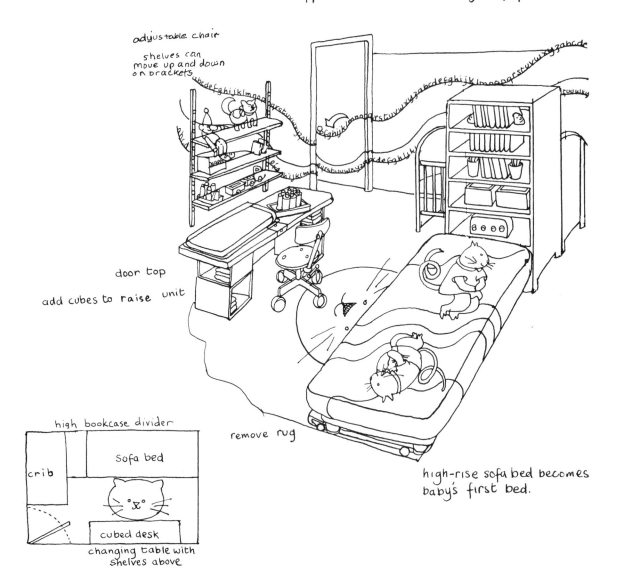

strippable, and removable baby wallpaper border.

adjustable chair

shelves can move up and down on brackets

door top

add cubes to raise unit

remove rug

high-rise sofa bed becomes baby's first bed.

high bookcase divider

crib

sofa bed

cubed desk

changing table with shelves above

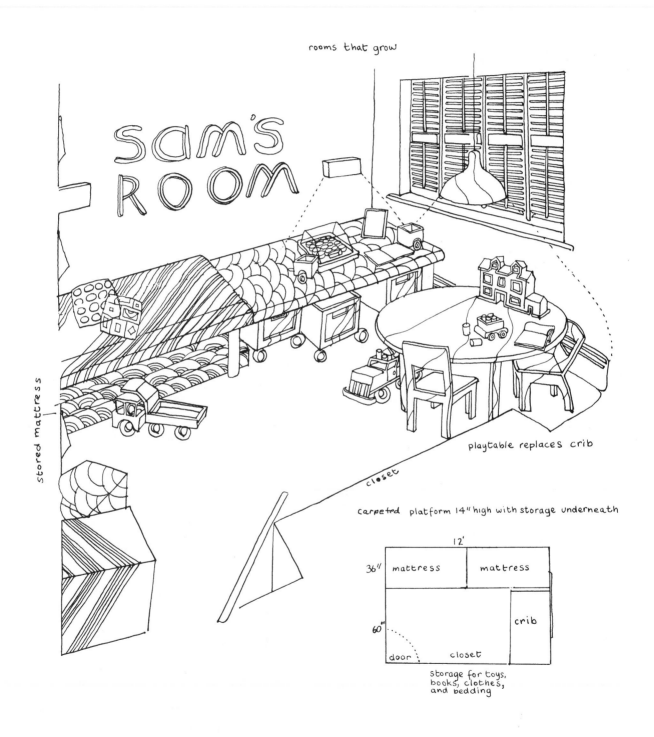

rooms that grow

sam's room

stored mattress

playtable replaces crib

closet

carpeted platform 14" high with storage underneath

12'

36" mattress | mattress

60" crib

door | closet

storage for toys, books, clothes, and bedding

Places to Put Nurseries

In very few households today, including those of the affluent, are parents not forced to adapt their homes to the arrival of a new baby, whether it is the first or the fifth. No matter if the baby will live in a large home or a compact condominium or a hopelessly undersized apartment, making the most of existing space is a universal imperative, for even those with extra space do not necessarily want their baby to take it over. Therefore your first challenge, which is discussed in this section, is to find and then plan a functional safe place for the nursery.

Places to Put Nurseries

> We put our son in an unused closet that was so small we thought that as he started to grow his legs would go through the wall.

Finding an unused, or underused, space for a baby is like playing a game of musical chairs. It's a question of rotation and reorganization. Other rooms must be designed to function round-the-clock, to be adaptable to more than one use. So first, forget about names that inhibit use—*dining* room, *family* room. Then think about location and sibling rivalry: who needs the most space and privacy, the baby, other children,

corner of livingroom curtained off

under a loft

you? Who will be displaced and how will they react? Should you give up your room? (See "Privacy in a Shared Space," page 40.)

Certain currently empty space that is not good for a baby—such as a loft (too high, although under a loft may work well); a garage and attic (too far away); a basement and porch (too damp); a hallway (too much traffic)—may be developed to free another more suitable area for a baby. In other words, if the baby moves into the den, create a new "den" in the corner of a kitchen or in an enclosed porch.

Conventions may have to yield and there will be some trade-offs. Because of space limitations you may now do your evening entertaining around the table in the dining room— where most of it takes place anyway. The baby could then occupy all or part of the living room, sharing it with you during the day.

If more than one child will share a nursery now or in the future, think about giving them your larger master bedroom with more play and floor space, so necessary for younger children. Then, later on, it can be divided into two bedrooms for older children.

A parent's place may not have to be fully enclosed at this stage, and it may make sense to have the baby and his "mess"

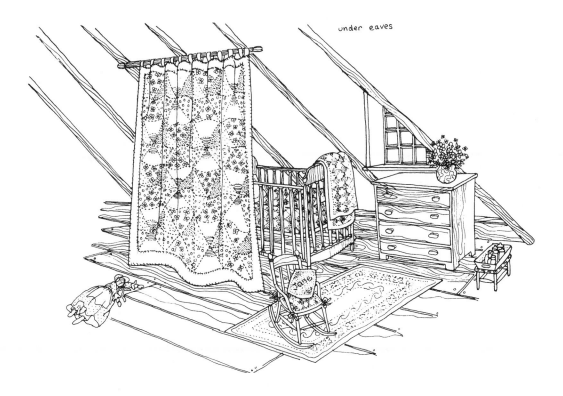

under eaves

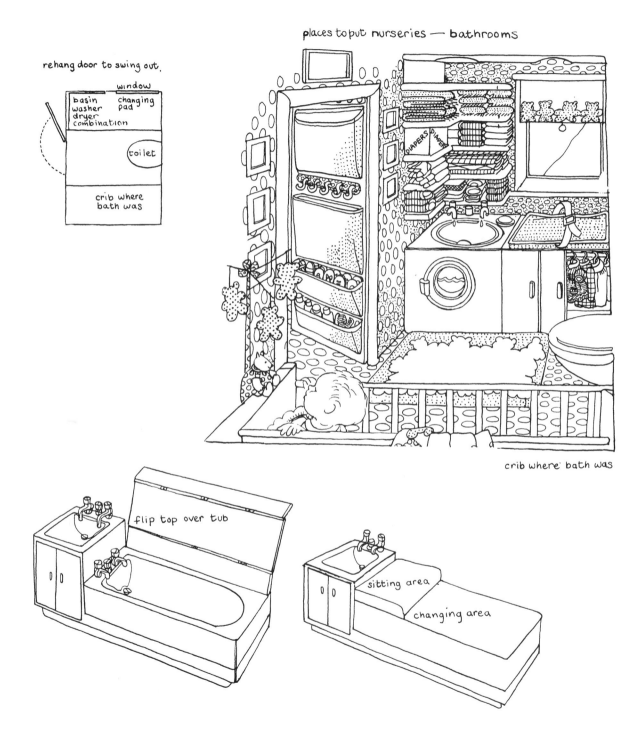

rehang door to swing out.

window

basin
washer
dryer
combination

changing pad

toilet

crib where bath was

places to put nurseries — bathrooms

DIAPERS

crib where bath was

flip top over tub

sitting area

changing area

hidden behind closed doors in a bedroom, even if it is the only one. After all, adults can occupy the rest of a home, while sleeping in the dining L, or a corner of the living room.

Use a corner of a den or living room for the newborn baby at first. Then when the baby is old enough to move in with his sibling, provide daytime play space for the older child in the newly freed den by adding concealed storage for toys. Make surfaces—floor, furniture tops—upholstery and washable.

A baby needs a full-time room, but occasional guests do not; put a sleep sofa in the living room for them and give the baby the guest room.

Set up a secondary play space near you by putting some low storage shelves for toys in a corner of the kitchen or home office, and providing your toddler with his own small table and chair, light and portable enough for him to carry from room to room.

Stretching Baby Space

There is no fixed formula for computing the amount of space necessary for bringing up a baby. You will be able to get by with less space without sacrificing convenience or aesthetics if you:

1. eliminate extraneous possessions
2. decorate to make an area appear to be larger than it is
3. select and arrange nursery furnishings to make optimum use of the space you do have.

So, if you are tight on space throw out or give away all that unused, unnecessary and underutilized equipment, furniture, and clothing you've been meaning to discard. Then try the following in the nursery as well as in spaces throughout your home.

Creating an Illusion of Spaciousness

With a soft, subtle monotone color scheme All baby blue, all beige floors, furniture, fabric. For interest, vary the textures and/or patterns and introduce another color in accessories, such as pictures and pillows. And to increase the effect, use the *same* color on floor or walls in adjacent areas, such as a hallway. Similarly, use one small-scaled patterned fabric on

striped walls, matching curtains

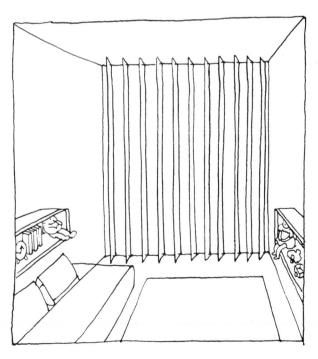

vertical blinds, low furniture

to raise a low ceiling

accent vertical molding

high bookcases to ceiling

everything: windows, upholstery, table, crib, and walls (stapled, glued, or shirred on rods [see pages 132–34]).

With uninterrupted furnishings For example, one unbroken expanse of floor covering with baseboard painted to match or built-ins that blend with the wall color will make the room seem bigger.

By eliminating extra accessories and furnishings For example, too many pictures, rugs, table and floor lamps, that a child can easily knock over anyway. Use wall or ceiling fixtures instead. Simplify window treatments and substitute neater and "lighter" shutters, shades, curtains, blinds, screens, panels of fabric (see pages 113–118) for heavy draperies or curtains that close a room "in" and take up space. And hide as much baby equipment as possible in the closet, under the bed, behind the bookcase, etc. (see pages 27–35).

By enlarging or adding a window A window or skylight added to a dark space will both brighten and "visually enlarge" it, as will a white or very light-colored floor or ceiling. Keep the window treatments spare and simple.

With murals and mirrors Fill a decorative frame with a mirror, put mirrors on one wall, on a closet or room door, in an alcove, perpendicular to windows (this doubles the size and amount of light). Just make sure the mirror does not catch too much sunlight and create too much glare, and hang securely. For full safety, a heavy mirror should be hung professionally.

By playing up vertical elements in a room with a low ceiling To make the ceiling seem higher than it really is, use a floor-to-ceiling window treatment, tall furniture such as bookcases that extend to the ceiling, and highlight vertical moldings with paint. Avoid low-hanging light fixtures that tend to pull the ceiling down—but *do* use low furniture and low-hung pictures.

Creating More Floor Space

By removing swinging doors Rehang a swinging door to open in a different direction, or change it to a sliding, pocket, or

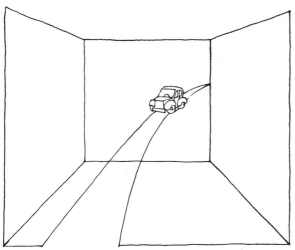

car and road painted on floor and wall of child's room.

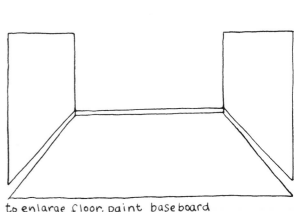

illusions of spaciousness

to enlarge floor, paint baseboard same color as floor

folding door; or remove it altogether to reclaim floor space. Warning: Folding doors can pinch little fingers.

By hanging furniture If there isn't room on the floor for it, practically everything but the crib, including the changing table, chests, cabinets, and shelves can be hung on the wall with angle brackets and bolts.

By buying multipurpose movable, folding, or stacking furniture Use chests of drawers, with a changing pad on top, instead of a separate chest and a separate changing table. Use a bookcase as a headboard, or a storage unit as a room divider. Use architectural features, that is, mantels and windowsills, instead of furniture for storage. Use small-scale furniture (but not so small that it isn't functional) and furniture that stacks, one chest on another. Whenever possible, says one parent, buy nursery equipment or furnishings that fold and/or have casters (wheels). And if furniture doesn't come with wheels, consider buying some and adding them.

substitute for furniture

an existing architectural feature, mantel with toys

How to Arrange
Juvenile Furniture
and Equipment

Small rooms or dwellings set the mind in the right path; large ones cause it to go astray.

LEONARDO DA VINCI

Comforting though it may be, a small space does present some overwhelming organizational problems that can be solved only with precise planning and measuring.

Study Clearances and Traffic Patterns

Furniture should fit so that it does not block access to windows, closets, and doors; so that drawers, doors, and sleep sofas and drop-downs have clearance space for opening; so that rockers can rock freely and not collide with walls. Leave enough space to be able to make the bed and to pull out a desk chair; and hang ceiling and wall light fixtures so you will not bump into them.

Place the Largest Piece of Furniture First

In a nursery this is probably the crib. It should not be under a window where a baby would be in a draft, and for your convenience you may want it nearer the door to the room. Place the changing table next to the crib, or near or in the bathroom and try to put storage furniture—chests, cabinets—in the closet.

Plan for Future Furnishings and Needs

As large an area as possible should be left free for crawling and playing, and space should be reserved for a bed, a play table, and chairs. Keep this in mind when buying or building in furniture.

How to Measure the Nursery and Make
a Floor Plan and Elevation

Measuring is a good way to study the details of the nursery area. And you'll probably discover while doing it that you can fit much more in than you thought. These measurements will also come in handy when the time comes to order furniture, paint, curtains, wallpaper, floor covering, etc.

Use a flexible metal tape, if possible, or a yardstick. Go around the perimeter of the room, noting on regular or graph paper: the swing of doors; permanent structures; such as pipes, beams, radiators, built-ins, an air conditioner and fireplace; sizes (measure outside edges) of windows and doors; and location of outlets. Calculate the distance, at the base of the walls, from corner to corner and from point to point; that is, the distance from the corner to the outside edge of the first permanent structure that you come to, such as a window, the width of the window, and on around the room. You now have a floor plan: a pattern of the room and its square footage (multiply one width by one length) of the floor and ceiling area. Since certain

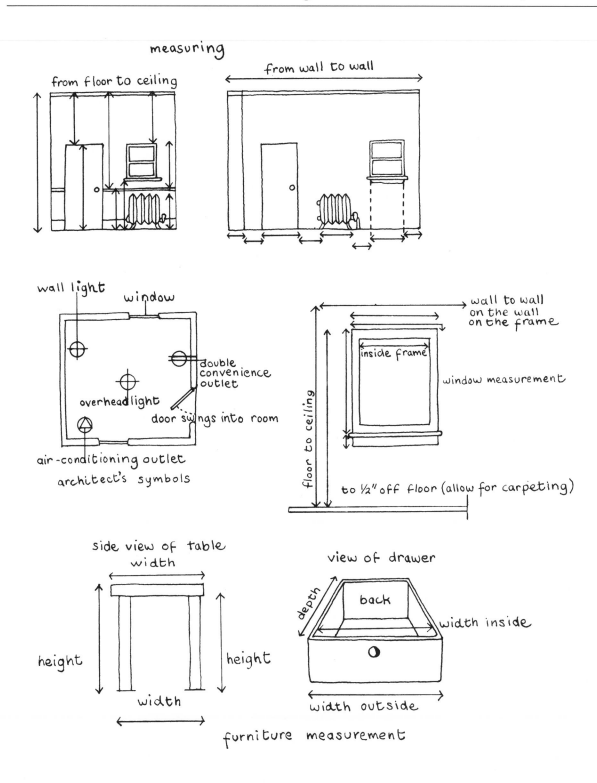

measuring

from floor to ceiling

from wall to wall

wall light

window

double convenience outlet

overhead light

door swings into room

air-conditioning outlet

architect's symbols

wall to wall
on the wall
on the frame

inside frame

window measurement

floor to ceiling

to ½" off floor (allow for carpeting)

side view of table

width

height

height

width

view of drawer

depth

back

width inside

width outside

furniture measurement

crucial details do not appear on the floor plan, such as the height of windows and doors, location of nonprotruding heating/cooling units and outlets, it is also advisable to make an elevation, or plan of the walls of a room. Do this in the same way, only now measure the distance from floor to ceiling.

If you are planning to build in furniture, or use furniture that fits a space so well that it looks built in (see page 32), elevation measurements are essential. If wall surfaces are uneven, measure widths more than once and use the smallest measurement.

Now transfer these measurements onto ½" graph paper (½" = 1') or onto the graph-paper grid on pages 157–166. Cut out the furniture templates on page 152, moving them around on the floor plan experimentally. Or try alternative arrangements by tracing the plan and then tracing the furniture templates in selected places right on the plan.

Standard Sizes of Furniture for the Nursery

The sizes below are for typical nursery furniture and show how much floor space each piece needs. Trace or cut out furniture templates, and arrange them on a graph-paper plan (scale ½" = 1') of the room. Add to these the cutouts for furniture, you already have. And instead of moving actual furniture, "life size" templates of furniture size can be cut from newspaper, placed in and moved around the room.

Note: Dimensions can vary slightly among manufacturers.

Crib: 30" × 54"
Porta Crib: 24"–27" × 42"
Crib and storage convertible modular units: 30" × 68"
    (see page 64)
Changing table: 15" × 34"
Chest of drawers: 24"–60"W × 16"–30"D
Bed: 39" × 75" (30"–33" must be custom ordered. A narrower alternative is a cot 24" × 72".)
Couch or sleep sofa: 30"–40"D × 60"–84"

Note: A 30"-wide bed or sturdy cot with a coil mattress is a big space-saver, fine for naps, thin adults, and most children, even grown-up ones. Thirty inches is a comfortable depth for sitting upright but a bit narrow for lying next to a child when reading.

A rocking chair, armchair or rocker-glider—about 18"–24" × 30"—should have arms to rest baby on. (If there is no room in the nursery for a rocking chair or glider, try to put one elsewhere in the house.)

Pull-up chair: 14"–24" sq.
Bookcase/shelves: 4"–18"D × 24", 30", 36", 42", 48", 54", 60", 66", 72"W
    (shelves can be as narrow as 4" for toiletries and small toys, 9"
    for standard-size books, 12" for games)
Play/work/changing counter: 18" × 34"
Play table: 24" × 24" × 36"
Round table: 24", 30", 36" diameter
Square table: 12" × 12", 15" × 15", 20" × 20", 24" × 24", 32" × 32"
    (card/game table)
Play-table chair: 12" × 12"

Typical Heights of Nursery Furniture

Full-size Furnishings:
Desk or chest: 26"–30" high
Chair or bed: 17"–18" high
Changing table: 36" high
Children's play/work surface/table and chairs:
Tables: 16"–22" high
Chairs: 9"–15" high

	Chair	Table
1½–2 years	9" high	16" high
3–5 years	10" high	18" high
4–5 years	12" high	20" high
5–7 years	15" high	22" high

Furniture should "fit" your child and grow as he does. Measurements will, of course, vary with the age and size of each child.

Storing Baby's Belongings

The number one space problem is not the baby but storage of baby paraphernalia.

The key to solving the problem of not enough space—and to living in some degree of organized chaos and harmony with children and their oversized equipment and overflowing possessions—is plenty of well-planned *designated* storage. One way to find it is to rearrange and exploit underutilized storage space that already exists someplace else in the house, throwing things out as you go. Another way is to look for new storage spaces in unusual spots—under stairs, on landings, over and on doors, down hallways, in corners and alcoves, and, of course, in attics, garages, and basements.

As you analyze your present and future storage needs, ask the following questions:

Can I put infrequently used items or outgrown baby clothes (store these in closed boxes labeled with size, i.e., 6–12 mos., 1–2 yrs.) in less accessible places, known as dead storage?

Where is storage located? Baby's toiletries and diaper pail should be kept convenient to baby's bath and changing area in the nursery, kitchen, or bathroom. It is only sensible to hang shelves low that you want easily accessible to your child, such as those for toys to encourage putting away, neat-

inexpensive metal shelves
60" high x 30" wide x 12" deep

wire baskets in steel frame
on wheels

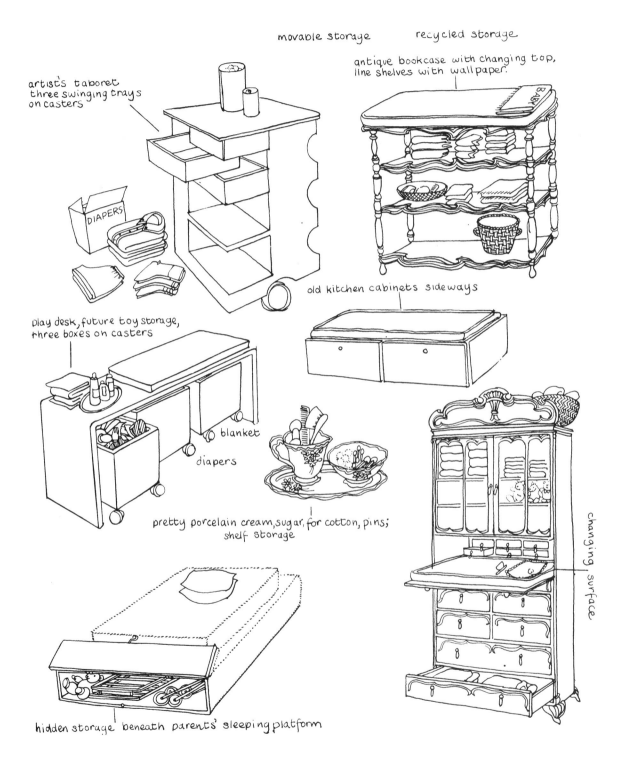

movable storage

recycled storage

artist's taboret
three swinging trays
on casters

antique bookcase with changing top,
line shelves with wallpaper.

DIAPERS

BABY

old kitchen cabinets sideways

play desk, future toy storage,
three boxes on casters

blanket

diapers

pretty porcelain cream, sugar, for cotton, pins;
shelf storage

changing surface

hidden storage beneath parents' sleeping platform

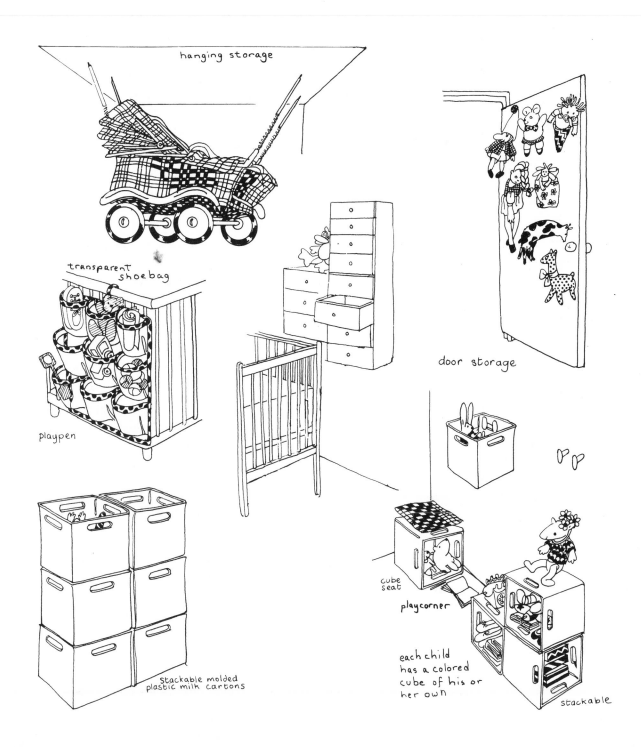

hanging storage

transparent shoebag

playpen

door storage

cube seat

playcorner

Stackable molded plastic milk cartons

each child has a colored cube of his or her own

stackable

ness, and self-sufficiency. On the other hand, shelves that hold items you do not want a baby to reach should be hung high.

Should the storage unit reveal or conceal? Open shelves can be sloppier, but they are supposed to be conducive to cleaning up: "There are the shelves—use them!"

Should the storage unit blend into the background of the room (thus seeming to take up less space) or should it be decorated and become a focal point of the room?

Overhead

Fitting a wide hall, floor to ceiling and overhead, with bookcases makes use of air space. So, look up (baby carriages *have* been seen hanging from the rafters). Shelves in the former den—now nursery—can then be freed for children's clothing and toys.

metal grids

In Closets

Disperse the contents of a linen closet directly to bedroom and bathroom shelves, or transfer the contents of a utility closet to a new closet or storage space in the basement or garage, and you'll find you have made room for all of the baby's equipment and, perhaps, even room for the baby.

Underneath

Look down! Shoving stuff under beds is nothing new, but doing it in a neat way is, with storage beds and beds on storage platforms.

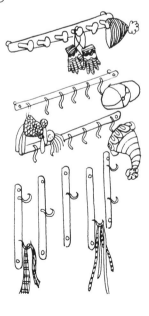

On Walls

Walls are often the most neglected but potentially most useful storage space in the house. By hanging furnishings you not only free the floor for playing (see pages 19–20) but you also fully utilize vertical space. Metal grids, shelves, storage cubes, chests, and cabinets can be on the wall reaching as high as you can. Hooks are particularly handy, and come in a variety of sizes from cup hooks to clothing hooks. There can never be enough of them *in* a nursery or, for that matter, *outside* it on adjacent hallway walls.

Wall storage can be restricted to one small area, to both sides of a closet or room door, or it can cover an entire wall. In fact, as we all know, the window wall is about the most wonderful storage and hiding place of all. Now, if you clear out the clutter currently behind the curtains and reorganize the wall in a more rational fashion with shallow wall-hung units and hooks, it will really work.

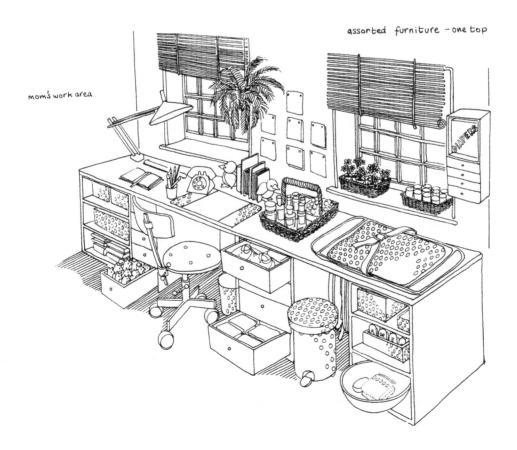

mom's work area

assorted furniture – one top

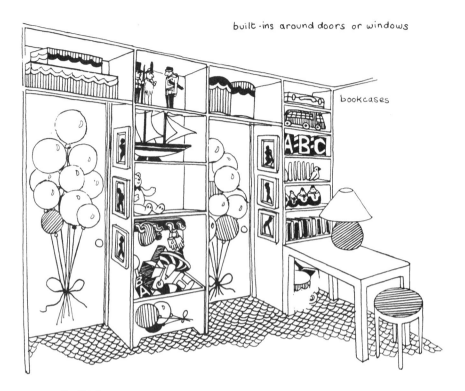

built-ins around doors or windows

bookcases

Built-Ins

And finally, there's the big question of built-ins. Should you have them; and where? It may not make economic sense in a temporary home or room unless they are built to be removed and used again. However, built-in storage is such a good space-saver: it looks trim and fits into nooks and crannies, making every inch count. In addition, built-ins can be less expensive than free-standing furniture, especially if they are constructed with a new or remodeled home. To compromise, use modular furniture that looks built in. These matching pieces are available, finished or unfinished, in a variety of sizes. Modular units can be placed next to each other to fill in a desired space—for example, wall to wall under a window—and can be rearranged to suit changing needs. To make nonmatching pieces pull together and look built in, paint or paper them the same as the walls and add one connecting top surface (see illustration) over two or more units that are the same height.

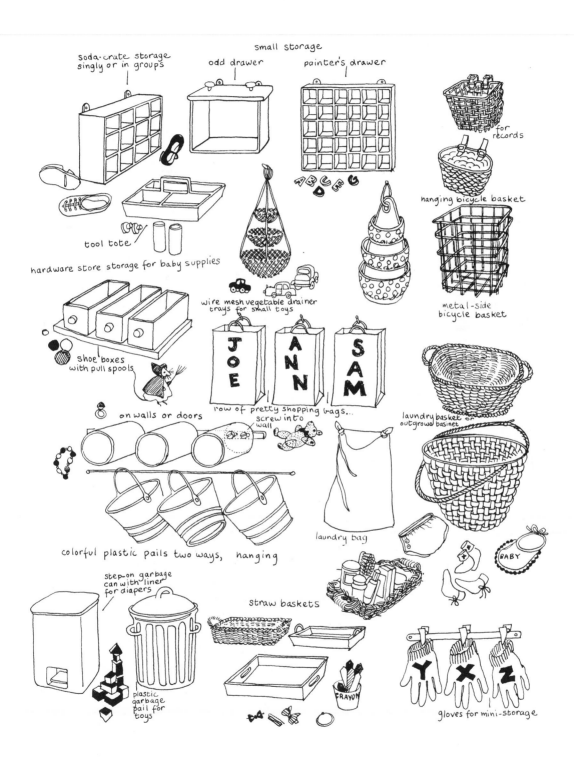

small storage

soda-crate storage singly or in groups

odd drawer

painter's drawer

for records

hanging bicycle basket

tool tote

hardware store storage for baby supplies

wire mesh vegetable drainer trays for small toys

metal-side bicycle basket

shoe boxes with pull spools

row of pretty shopping bags... screw into wall

on walls or doors

laundry basket or outgrown basinet

colorful plastic pails two ways, hanging

laundry bag

BABY

step-on garbage can with liner for diapers

straw baskets

plastic garbage pail for toys

CRAYON

gloves for mini-storage

JOE ANN SAM

Y X Z

stacking furniture

painted like wall

Functional Storage Furniture

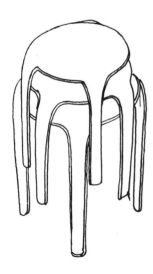

Regardless of whether it was intended, plenty of furniture all over the house will be pressed into service to hold baby belongings. Therefore all newly acquired furniture should have maximum "storability." Tables, chests, desks, coffee and end tables should not have spindly little legs that waste space. Instead they should have many shelves or drawers. (Children's drawers should not be so large that they become stuffed, and then too hard and heavy to open.)

Shelves should be selected with the objects to be stored in mind: deep enough (12") for bulky games, but not so deep that smaller toys are forever lost in the back. Shallow shelves as narrow as 9" can be used for books, as shallow as 4" for tiny toys and toiletries. "Shallow shelves are easier to keep neat than deep shelves," says one experienced parent. Walls of shelves can be installed on spring or suspension poles or on standards and brackets. They can be free-standing metal, wood, or plastic shelf units or "bookcases."

Stackable, molded plastic or wood storage cubes These provide very versatile storage. Usually about 16" square (like milk cartons) with casters, they can be put under the crib, under play tables, stacked (but not so high that they might be pulled over), or hung on walls. Upside down, they are a table and with a pillow added, a stool. Cubes can be used for blankets and diapers now, for toys later on, and eventually for books and records.

New or old furniture originally intended for another use Pieces can often be adapted, adding not only storage space but style to the nursery. For example, an artist's taboret usually holds paints, papers, and brushes, but it is perfect for storing diapers, small clothing, toiletries. Since it has wheels, it can be pushed around from bath to crib, even from room to room, wherever needed. A serving wagon could "serve" in a similar way, and a white-elephant breakfront or secretary, too ugly and cumbersome for another area, might be marvelous for multipurpose storage and baby-changing.

Decorative "antique" oldies Baby clothes will look pretty hanging out in full view on a Victorian wicker, wrought iron, or wood hall rack with hooks, shelves, and mirror. Old bookcases and chests can be charming, as can old oak refrigerators, dentist and doctor's cabinets. Use odd pieces of china, such as a sugar bowl or coffee cup, for cotton balls and old tins for tiny toys. Restore or restain a finish that is too far gone (see "Rejuvenating Old Furniture," pages 109–12) and repaint an old (possibly lead-paint) finish that is flaking; sand this first.

Small storage from the hardware store A variety of household items can be adapted for nursery storage, such as bread boxes and mailboxes, garbage pails, totes for tools and silverware, shower caddies and trays. (I like to put trays on shelves and pull them out as needed. That way nothing gets lost in the back. Small items are also clearly visible spread on a shallow surface and not hidden at the bottom of a deep pile.)

Closets for Children

Closet space is a costly and a rare commodity, even in larger homes. So if you are lucky enough to have a closet for your child, make the most of it. The average closet is 2' deep and from 3' to 6' wide, with one high pole and one distant shelf, good only for bulky bedding and dead storage. Additions can be made by leaving the pole and shelf intact or moving or removing them and completely redesigning the closet. Take into account both the future contents of the closet and accessibility for you and your child. (You will want him to reach certain hooks and shelves to put things away but not reach others.) Plan a closet carefully and it can hold all nursery storage and some of yours as well.

Shelves Assorted sizes can fill every extra inch. Narrow ones can be used on doors and on inside side walls. Larger shelves can be installed the full width or height of the closet.

Cabinets or chest of drawers These can be stood side by side or be stacked, for maximum storage, from closet floor upward all the way to the ceiling.

Poles You may not need these at all. Chests and shelves are more practical for early years—until age 5 or 6 most clothing lies flat—and for boys of all ages, whose total hanging wardrobe often consists of two jackets. If you do plan to move or install poles, keep in mind that a three-to five-year-old can

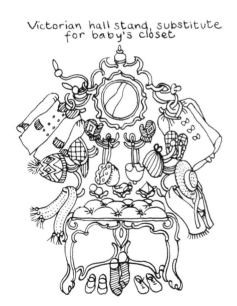

Victorian hall stand, substitute for baby's closet

changing closet

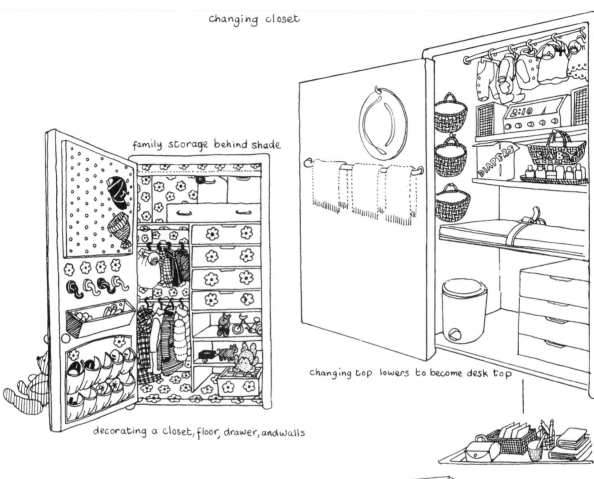

family storage behind shade

decorating a closet, floor, drawer, and walls

changing top lowers to become desk top

multi-use storage unit, extra shelves
in between box cases, rods for towels and clothing

reach a pole about 30" from the floor, an older child about 45". One pole placed near the very top of the closet that only you can reach may be useful. This could be a spring rod installed over the existing upper shelf for hanging baby clothes.

Hooks Devices such as hooks, special hangers, racks, boxes, kits with necessary parts and hardware for redesigning closet space, and wonderful gadgets aimed at expanding closet space are available at hardware stores and closet shops.

Making a Changing Closet

With precise planning it may be possible to use the closet for dressing by adding a semipermanent changing shelf (this can be removed eventually or be lowered and become a writing/working surface) or a movable changing wagon. Small shelves for diapers, clothing, linens, dressing toiletries, and a diaper pail can then be added above and below.

Door Storage

On a swinging door you can hang bulletin and pegboard, metal grids, hooks, racks, narrow shelves, a mirror, a shoe bag (for toiletries and toys rather than shoes). Almost anything can be put up if it is shallow enough for the door to close easily without colliding with the other contents of the closet.

Different Kinds of Doors

For greater visibility, or because you need the space that a swinging door takes up, you may want to replace it with a folding (bifold) or sliding door. (It's a trade-off, though, because you'll lose all that good door storage space.) If you would like your young child to open the closet easily, lower the doorknob. To make a swinging closet door lighter, cut it in half horizontally (you'll need two knobs now), or remove the door altogether—nice only if the closet area is neat and decorated to complement the room. (Keep the door so that it can be reinstalled if you have second thoughts, or for the future when the closet becomes more of a disaster than a decorating asset.)

Privacy in a Shared Space

When can you put two children together? How can you divide a space, particularly a small space, so that each child has his own working, playing, sleeping, and storage space? Should there be a separate sleeping area and a common play/work area, or vice versa? Should a room be permanently divided with a solid wall for total sound and light insulation, or should the spaces be separated with a movable partition, such as folding screens? Or should there be just a suggestion of separation made by a change of flooring or a furniture divider? The answers depend, of course, upon the special situation of each family, and the gender and ages of the children, as well as the prejudices of parents that arise out of their own childhood experience: "My sister and I shared a room until we were eighteen and we were, and still are, very close." "My sister and I shared a room and fought constantly. I would never put my children in the same room. In fact, I would put them as far away from one another as possible."

Younger children seem to like the comfort of company and sharing with a new sibling as long as Baby is quiet, old enough not to be up all night crying, and there is someplace else in the house to play when Baby is sleeping. For them it is more important to have separate and special storage for their possessions rather than a private place to sleep. Around age

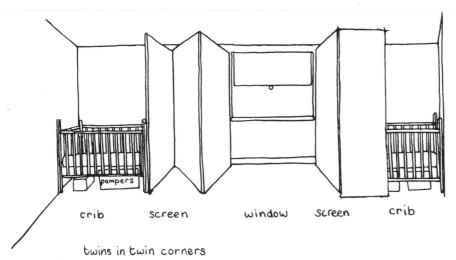

crib   screen   window   screen   crib

twins in twin corners

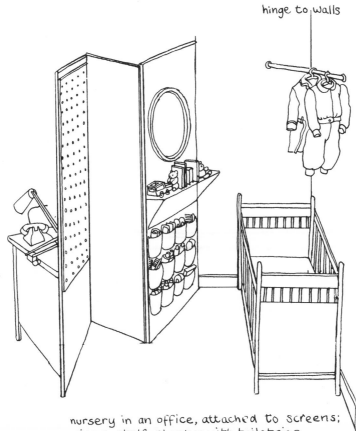

hinge to walls

nursery in an office, attached to screens;
mirror, shelf, shoebag with toiletries,
pegboard.

room dividers

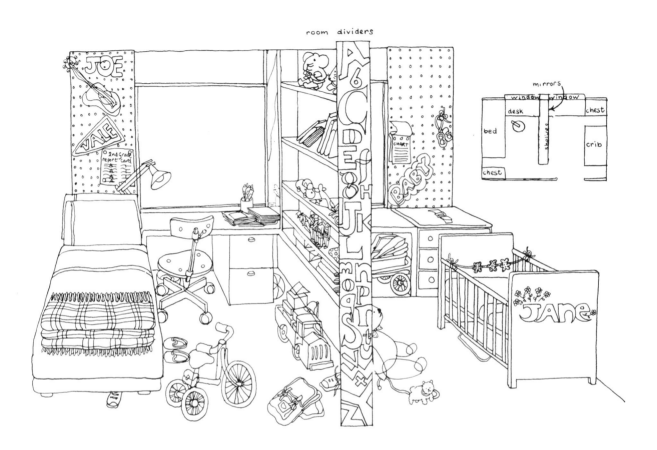

mirrors

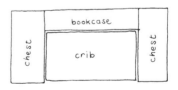

furniture

six, children start to have a greater sense of themselves and an accompanying need for privacy and a separate living space, no matter how small. By age seven or eight, children of different sexes should definitely be separated in some way.

Given the reality of today's very cramped living quarters, it is more than possible that parents, as well as siblings, will be sharing a room. While it is quite common for a baby to remain in a parent's room—even "bonding" in their bed—for the first few months, it is generally recommended that for everyone's privacy, the baby should move out as soon as possible and no later than six months.

Dividing Space

Note: The following ideas apply to dividing any area to create a separate "room."

Ideally, each child should have a permanently split, separate, but equal room with a window giving light and air, artificial light, a heating unit, a closet (nice, but not necessary), and a door into it from a hallway, not from another room.

To lessen the claustrophobic feeling in a now smaller space and for more light, install a transparent material on the upper part of the partition, or an interior window (if you are concerned about broken glass, use one made of shatterproof plastic panes).

Sometimes a permanent wall or divider is not desirable because it is too expensive, the arrangement is temporary, or the space doesn't lend itself to division because access from a common area is not possible or windows are placed so that they cannot be shared. In such a case a movable partition may provide both a degree of privacy and the required light (but not sound) control. This could be a folding accordion divider or a curtain divider or sliding panels of fabric, wood, or shoji (Japanese rice paper), installed on a ceiling track. (Needless to say, fragile material such as the Japanese rice paper used to make shoji screens may be fine in an infant's room, but it will not withstand your more athletic two-year-old's activities.)

To allow light and air circulation, another alternative may be to build a solid wall perpendicular to the width or length of the room a foot or two lower than the ceiling. A one-track lighting system can then, by directing the individual lights, service both sides.

In many cases a wall or wall-like structure is not necessary and a room or space can be "divided" by:

Placing furniture to establish two distinct areas: low chests, couches, beds, desks or high bookcases, cabinets or free-standing closets. (Just be sure high and heavy furniture is bolted to the floor and/or wall.)

Using two kinds of color for wall or floor covering. If stor-

breakfront/room divider
separate

age is shared, amplify this idea by painting each child's drawers the color of his rug or walls.

Buying or making room dividers.

Temporary Room Dividers

Folding screens useful in shielding a baby from light and drafts, screens are also very versatile and economical room dividers. They can enclose a corner or an alcove and create a totally separate space. They can also be easily removed when no longer needed. Screens can be any height or weight. (If a screen is

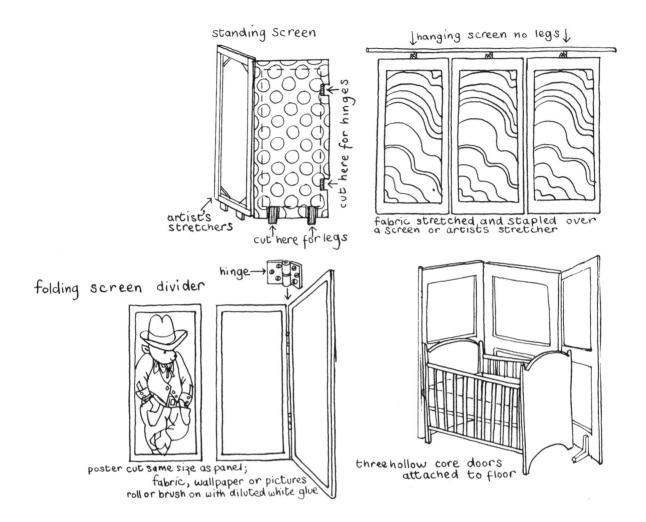

standing screen

artist's stretchers

cut here for hinges

cut here for legs

↓hanging screen no legs↓

fabric stretched and stapled over a screen or artists stretcher

folding screen divider

hinge→

poster cut same size as panel; fabric, wallpaper or pictures roll or brush on with diluted white glue

three hollow core doors attached to floor

lightweight and likely to be tipped over, place it next to a wall or a solid piece of furniture.) They can be free-standing or attached on one side to a wall with hinges; or hung from the ceiling with hooks. Usually made of three or more panels in a variety of material including wood, shoji, rattan, metal, leather, or plastic, they can also be covered with posters, wound with ribbons, painted, papered, or otherwise decorated (do the two sides differently). A very sturdy screen—such as made from hollow-core doors—can be functional as well as decorative with lightweight baby storage (bulletin or pegboard, small shelves, hooks, and a mirror) hung on the

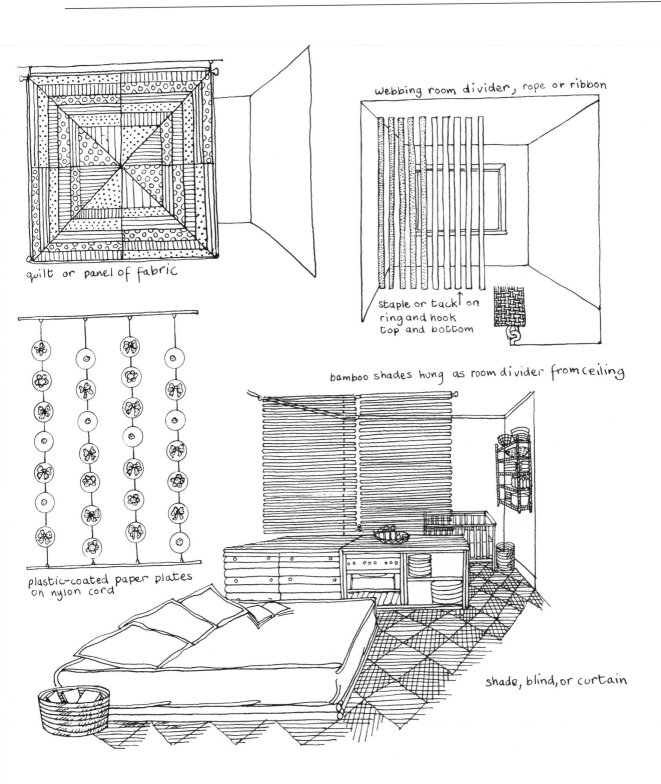

quilt or panel of fabric

webbing room divider, rope or ribbon

staple or tack on
ring and hook
top and bottom

bamboo shades hung as room divider from ceiling

plastic-coated paper plates
on nylon cord

shade, blind, or curtain

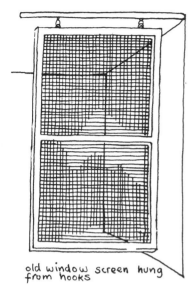

old window screen hung
from hooks

baby's side. Screens with double-hinged panels can be moved backward and forward into several configurations. And even if a screen is not needed to divide a space, it can be a nice decorative addition, especially if it is a pretty, old one.

Ready-made dividers A variety of types are sold in stores and are expressly intended to be room dividers. They may be spring-pole units fitted with cabinets, or shelves, or solid panels. But there are other items available, especially in the drapery department, that will divide a space as well, such as bead, bamboo, or fabric curtains; venetian horizontal or vertical blinds; and window shades. These can be installed on a ceiling rod, or on hooks in the ceiling (and floor if necessary), and can extend from ceiling to floor or to the top of a furniture divider.

Homemade dividers These can be made from almost anything, including ladders, old—and clean—window screens, but not the kitchen sink. Panels of fabric are decorative as dividers and easy to work with: simply attach rods the same length as the fabric panel to the top and bottom of a panel or stretch the fabric on artist's stretchers, available in large sizes. Heavy fabric such as canvas, or a quilt or a large picture can often be hung without these supports. Strips of upholstery material, such as webbing or ribbon, can be installed from floor to ceiling, and cord, wool, string, or ribbon can be wound around or woven through decorative panels or spring poles.

Measure the width and height of the area to be covered carefully and allow a few extra inches of material at each end (top and bottom) for attaching the divider. Attach dividers made from thin, lightweight material with tacks or staples, heavier ones with cup hooks or heavier hooks in the ceiling beam, putting screw eyes or clip or sew-on rings or curtain pins on the divider. To test if a ceiling-hung divider is straight, drop a plumb line (a scissor tied to a string will do) from the ceiling to line up parallel with the side of the divider.

glue two plates back-
to-back on string. tie
to screw eye on floor.

A Safe Space: Childproofing Your Home

It is easy to underestimate the future strength and agility of today's gentle bundle, who will very quickly be tomorrow's terror and terrible two-year-old. So prepare yourself—and your home. All at once it seems, he'll roll over, reach out, and stand up, and you may not be there the first time he does it.

A safe home will ease your anxiety and encourage your baby's activity; so the whole house, not just the nursery, should be checked and made childproof.

Cover furnishings that have sharp corners or edges with foam tape, or remove altogether. Substitute soft-edged furnishings, such as a foam couch without a frame.

Remove anything breakable, including furnishings with small pieces that can break off and be eaten, and coffee table bric-a-brac. Store or hide cigarette lighters, sewing equipment, and small objects such as paper clips and coins.

Cover unused outlets and extension-cord receptacles with outlet covers. Conceal electrical cords (sometimes furniture can be moved in front of outlets and to cover cords).

Substitute for furniture with hard edges and surfaces (marble, metal) furniture that has rounded edges (i.e., a round coffee table) and a softer surface—wood, or better still, all-foam furniture (chairs, love seats, table/stool combinations).

Foam furniture is also lightweight and usually washable. Cover remaining furnishings that have obtrusive hardware and sharp corners with foam tape or the special protective covering that is sold for this purpose. And if you have any old upholstered pieces, be sure to check the condition of the stuffing. Babies love to pull it out and eat it!

Secure furniture, and firmly install shelves and anything else that hangs. Standing furniture should be solid on the floor so it cannot topple over. For extra security, place tall furniture against a wall, and attach it to the floor or wall.

Plan storage so that harmful things (medicine, cleaning agents, matches) are high and out of reach. For example, rearrange kitchen storage to remove sharp tools, electrical appliances, and cleaning supplies from bottom kitchen drawers or cabinets. Substitute safe plastic tools and toys a baby can reach and safely play with.

Even better, install locked cabinets. Put latches (or strong strapping tape) on low-level drawers and cabinets, particularly in the bedroom, kitchen, workshop, and garage. Also, restrict access to exposed heaters, radiators, pipes, and fireplaces. Use (and maintain—i.e., check batteries, etc.) a smoke alarm, carbon monoxide detector, fire extinguisher, and a baby monitor.

Remove or push back from shelf or chest easily overturned objects such as lamps and telephone, and don't leave cords dangerously dangling within a baby arm's reach. To prevent a child's tangling, keep cords and strings attached to hanging toys away from the crib. Once the baby starts to crawl, tie up drapery or blind cords, or wind them around a plastic cord shortener; or just tape up or rubber-band any excess cord. Loops in blinds and cords pose a strangulation hazard to children. Safety tassels are available free of charge at window covering retailers. Call 800-506-4636 to order free tassels.

Purchase child-protection products from a hardware or baby store or from the excellent catalog and web site sources at the back of this book. Buy special locking devices and latches for cabinets, drawers, kitchen appliances, and bath-

safety devices

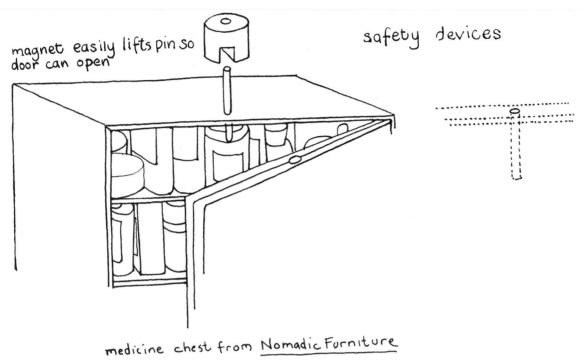

magnet easily lifts pin so door can open

medicine chest from <u>Nomadic Furniture</u>

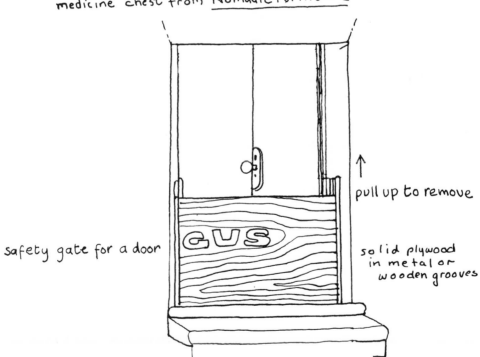

safety gate for a door

pull up to remove

solid plywood in metal or wooden grooves

GUS

room fixtures. These include toilet locks, spout covers, and a device that limits the heat of water. Also, purchase outlet covers, protective covering for corners and edges, and window guards (to keep windows safe you can also make sure to open them only from the top, or install locks or nails in the frame so bottom windows open only a few inches). Finally, install folding gates made for babies (or pets) to block doorless doorways and stairs at the top *and* the bottom and any other potentially hazardous areas and passageways (hardware mounted gates provide greater security than pressure mounted).

Most parents recommend mesh or gates of solid wood over the trellis-like wooden folding gates that an agile child can climb over or get caught in and strangle. (Note: A child can also get stuck between stair balusters or fall through those that are more than four inches apart.)

Look out for lead in surfaces such as walls or furniture painted before 1960. Chipped and peeling walls should be repainted or papered. Lead dust can also be released in the course of restoring an older home. Lead poisoning can be particularly hazardous to children.

Avoid furnishings with free-falling hinged lids, such as toy chests or trunks, as they can present a serious strangulation hazard. (Toy chests sold today are supposed to have safe hinges.) Do not use a large storage container in which a child may be trapped and suffocate. If a toy chest lid does not have a support, either add one or remove the lid altogether. Alternatively, use storage with sliding doors or open bins and shelves, which are more convenient for toys anyway. (Consult *Consumer Reports* or the Consumer Product Safety Commission web site, www.cpsc.gov, for the most up-to-date information on recalls.)

Though mentioned a few paragraphs back, it cannot be repeated often enough: nursery furniture should have *smooth surfaces and rounded edges or points,* and restraining straps where appropriate—that is, in high chairs, strollers, carriers, changing tables, and infant car seats. Some juvenile equipment, such as cribs and high chairs, now bears a label of safety approval from the Juvenile Products Manufacturers Association (JPMA).

Then, play baby—crawl around the house to spot any other potential hazards, such as a floor plant your child might pull over or eat! For your own safety at night in the nursery, install a nightlight or a lamp with a dimmer. Finally, if the nursery is newly painted, papered, or carpeted, open the windows to ventilate the room a few days before the baby moves in.

Furnishing Baby Space

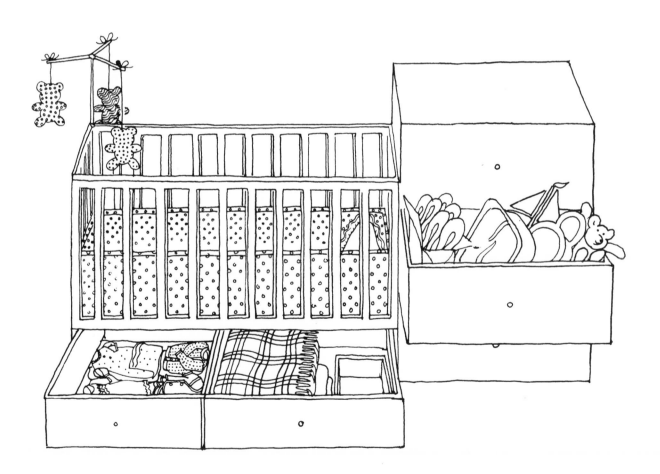

Necessities for the Nursery

Now that you have found a space for your baby, you must figure out what equipment and furnishings you will need throughout the various stages of babyhood. With cramped space, selectivity in what you purchase is the key.

Where to start? And when? You have probably thought about furnishing and equipping the nursery, on and off. When the baby becomes more of a reality, generally about three months before it is due, many parents suddenly realize they must "do something"—and they really *must* if baby furnishings such as a crib are special-ordered—stores need this much time. Casual chats with family and friends and browsing in baby boutiques should be translated into action and some serious buying, or at least ordering. Practically everything that is needed in the first months can be selected ahead of time. And if you are superstitious, delivery can be delayed until after the baby is born.

Parents cry that their first visit to a baby store or department was "anxiety-producing," and no wonder! So many things are manufactured for babies today that you can develop guilt feelings for child neglect if you do not end up owning many of them. In deciding what to purchase, keep your storage facilities in mind. And try to select equipment that is flexible, folding, mobile, and if possible, multipurpose.

"Practically everything seems to double—or triple—as something else," said one mother who discovered that she could use her playpen as a crib and her car seat as an infant seat.

Equipment Checklist

Bassinet
Baths & bathing aids
Beds (youth, bunk, cot, twin)
Booster seats
Bouncy seat
Bottle warmers
Breast shields & pumps
Car seats
Car-seat covers
Carriages
Carriage bags
Carriage & crib nets, wind train shields
Carriers (backpack or front carrier; cloth and molded)
Chairs (arm, rocking, glider)
Changing pad (can be placed on a dresser or other flat surface to make an instant dressing table; contoured mats keep the baby from rolling)
Chests & changing table or dressing surface
Cradles
Cribs
Diaper & formula bags
Diaper pails (with deodorizers except Genie)
Exersaucer
Execusor
Feeding accessories
Feeding tables, hot plates
Gates & guard rails
Hampers
Hangers (children's clothes)
High chairs (reclinable recommended)
Humidifier
Lamps
Liners (bassinet and basket)
Mattresses (crib and carriage)
Mobiles (crib)
Monitor/Intercom
Play yards or "Pack and Play"
Safety Products
Scales
Step stools
Sterilizers
Strollers
Stroller seat covers
Swings (indoor)
Switch plates
Table and chair sets
Thermometer (rooms)
Pacifiers & teething aids
Toys & toy chest

You will also need the following:

BEDDING CHECKLIST

Fitted bottom sheets (top sheets are not necessary) for crib, carriage, or bassinet (3 each—white knit are nice, or those that match the room).

Cotton or cotton flannel fitted crib and carriage sheets

Blankets. For general purposes and for carrying: 3 lightweight thermal and 4 washable cotton receiving blankets. For the crib a medium-weight blanket and for the carriage a heavy blanket. Quilts are attractive. (A lovely quilt could hang on the nursery wall.) Comforters (and other soft bedding like pillows) are safer in an older (2+) child's bed.

Cloth diapers (even if not worn, they are handy for holding, feeding,
 protection on dressing area, etc.)
Burping cloths
Bedding accessories: *bumpers*, crib skirt, pillow sham

To make up carriage or crib:
1st: waterproof sheet and/or mattress cover (if mattress is not plastic)
2nd: fitted sheet
3rd: small waterproof sheet or pad with cover (these are also useful
 on the top of the changing table)

BATH AND TOILETRY CHECKLIST
Soft terry washcloths (2–4), bath and/or hooded towels (3), plus a few
 small towels (these can be cut up from some old large towels)
Mild soap (keep it in a dish that is big enough to hold a little water
 for a sponge bath)
Cotton balls and cotton swabs (in a covered jar)
Baby lotion, oil, shampoo (no-tear type)
Cornstarch powder
Wipes (wipe-warmer optional)
Diaper-rash cream, A&D ointment, Vaseline
Rubbing alcohol (to cleanse umbilical cord stump)
Diaper pins with plastic heads (if you use cloth diapers), hairbrush,
 and fine-toothed comb, pacifier
Baby fingernail scissors
For colds: a vaporizer, nasal aspirator, dropper, rectal thermometer,
 digital ear thermometer
For formula: bottles, nipples, tongs, bottle brush, nipple brush
Hand sterilizer (i.e., Purell)

Toiletries should be easily accessible, and if necessary,
easy to carry around. They can be stored on the tray provided
with some changing tables—or on a separate tray, in baskets,
plastic or pretty cardboard shoeboxes, or any other suitable
container you may have around the house.

Basic Rules for Buying Baby Furnishings

Don't limit your shopping to the baby department. Juvenile-furniture stores and baby departments have become responsive to the needs of today's families and now stock more practical, classic, and compact furniture and fewer delicate strictly-baby pieces. However, you still owe it to yourself to explore all other sources:

For new furnishings Furniture stores that specialize in children's—as opposed to nursery—furniture; the regular furniture stores, and the furniture department in department stores, specialty import stores, mail-order houses, the five-and-dime, and hardware stores.

For secondhand furnishings Yard and garage sales, antique shows, auctions and church bazaars, merchandise offerings in newspapers (and put your own want ad in).

Don't forget hand-me-downs Whether you borrow them or use furnishings you already own but have forgotten about—such as that ugly oversized armoire you inherited and immediately relegated to the garage—these can come in handy. Painted over, they might provide more than enough storage for several babies.

Consider the needs of an older child now Work backward: your child will need a bed and a variety of surfaces for playing, working, and, perhaps eating, such as a desk or table (near natural light), a platform, the floor and wall. She will also need chairs and other places to sit—a floor pillow, a banquette or a window seat; general and task lighting (work, play, reading); and lots of storage for clothes, books, toys, records, on shelves, in chests or cabinets, or in one large unit that incorporates all of these, plus a desk.

Select flexible furniture Whatever you choose should grow with your child: shelves, tables, desks and chairs that can be lowered and raised, or tables with parts that can be added on to make the legs higher; pieces that can be adapted to other uses: for example, a crib that can become a youth bed or a settee, or headboards for a bed; a play table that can become an end table or coffee table. Steer clear of matching suites of nursery furniture—these are uninteresting and quickly outgrown.

Buy the best quality furniture for the price It should be sturdy, safe (no sharp corners and edges), and washable—a lacquer paint or plastic laminate finish all over, or at least on top surfaces, is excellent. To check for structural soundness in a cabinet or chest, push and pull it, see if the joints creak, if the drawers and doors open and close easily and smoothly. Better pieces should have good hardware, a smooth evenly matched finish, center or side guides for drawers, and dovetailed drawers with reinforcing blocks in corners underneath.

On the other hand, since part of the fun of fixing up a nursery is that it will change as your baby does, there's nothing wrong in buying a frivolous-and-frankly-junky piece intended just for now, and not forever (see also Functional Storage Furniture, pages 34–35).

Make sure you have recourse when buying baby equipment, especially a crib (see page 64), and take certain precautions for its return if something goes wrong. Find out about the store's policy (will you need the original box, for instance?). Keep your sales check (with date, price, model number, description of item) as long as you think necessary.

If the equipment is faulty, do not be shy about notifying the store immediately. Speak to the manager or someone with the authority to replace it. If you get no satisfaction, you may have to write or call the president of the manufacturing company; give him the information and ask for immediate action by a specified date. (Make copies of your letters in case you have to proceed further to the U.S. Consumer Product Safety Commission in Washington, for example.) Follow up on it if you haven't heard by this date by calling the president or the consumer-relations director of the company. Manufacturers of baby equipment seem genuinely responsive, so you may even succeed without shouting too much.

A Place to Sleep: Cribs and Cradles, Bunks and Beds

A great many parents, I find, seem concerned from the start about where their baby will sleep. The answer to that initial question is that he can sleep anywhere there is room.

A baby does not need a bassinet, a fancy crib, or some other sleeping place designed (presumably) for his needs. If you want him to have one, fine. But he doesn't need it, either physically or psychologically. What he needs is simply something that is big enough for him and that will shield him from danger. For a newborn baby, a dresser drawer or a large old-fashioned wicker laundry basket functions beautifully. To be perfectly safe, it should be at least eight inches deep—and should be placed where an older sibling does not have ready access to it.

VIRGINIA POMERANZ, M.D.,
The First Five Years: A Relaxed Approach to Child Care

Short-Term Sleeping Places for Infants

If for one reason or another—budget, space restrictions, or family logistics (an older child has not quite outgrown it)— your new baby does not have a crib, do not feel guilty. For the first two or three months babies seem to prefer a confined space. You can take advantage of this by using one of the following, fitted with a small mattress, a piece of foam, or a soft folded blanket.

Cradles A universal favorite of babies because of the "womb-like" rocking motion, they are also loved by parents because they can be placed next to their bed and rocked almost in one's sleep with an extended foot or a long string. Because a cradle has very low sides, as soon as the baby starts to pull himself up, switch to a crib (and use the cradle for magazines and/or toys).

Moses basket (carrying or laundry basket), carrier cradle, drawer, or sturdy box All should be at least 30" to 36" long and rest either on the floor or on a very solid table, or in the crib itself (to make the crib cozier, at first). Baskets should be smooth and splinterless. Like a bassinet, they can be painted and covered with a fabric, such as lace or muslin, and ribbons.

Carriages These make a very convenient first or supplementary bed. They can be wheeled from room to room, kept in your bedroom, and rocked as you rest in bed.

Bassinets Always a sentimental favorite. Because they are outgrown before they are worn out, they are often passed down for generations in a family, with new and different ribbons and skirts added for each newborn.

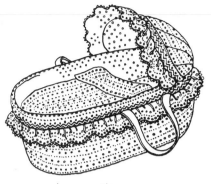

carrier cradle

bassinet:
ribbons that match room
woven through bassinet
or basket.

Cribs

This somewhat jaillike environment is your baby's world where he will spend most of his first two years—usually until he is walking and toilet-trained—a good part of the time asleep. It is the single most important baby furnishing. Unless you are inheriting, borrowing, or renting one, a crib plus mattress is expensive, so it should be chosen with care.

There's a crib style for every period. Country American, Victorian, modern are manufactured in several materials, including wood and metal. Full-size cribs are approximately 30" × 54" (size varies, particularly if the crib is imported) and on average cost from $100 to $200 without the mattress. Some cribs have one or two movable sides or one folding drop front or fixed sides. Some convert to "training beds," junior beds, or loveseats when the sides are removed. These sides can later be the headboards for regular beds (one manufacturer has gone so far as to suggest that crib sides be used first for newlyweds, until they need a crib).

A relative newcomer to the nursery is a compact convertible modular unit with crib that seems to combine all necessary furniture for sleeping, storage, and dressing. You will, of course, need both more floor space (approximately 30" × 68") and more cash (approximately $500 to $1000) in order to purchase one of these units; on the other hand, the life of the furnishings is substantially lengthened, since the unit separates into several of the following, depending upon the model: chests, cabinets, a desk, and a regular junior or trundle bed.

One interesting note: several years ago a group of German scientists found that infants raised in transparent Plexiglas cribs that afforded them total visibility were better able to see and react to what was going on. As a result, their mental development was much faster and by 18 months they were measurably more intelligent than the control group raised in traditional cribs. Applying this argument, you may want to consider a crib with open or slatted endboards. Even if the baby's IQ is not affected, you will have a clear view of baby and what baby is doing.

Safety Standards for Cribs

The federal government, through the Consumer Product Safety Commission (CPSC), has imposed several regulations on the manufacturers of cribs. Even so, accidents do happen, as illus-

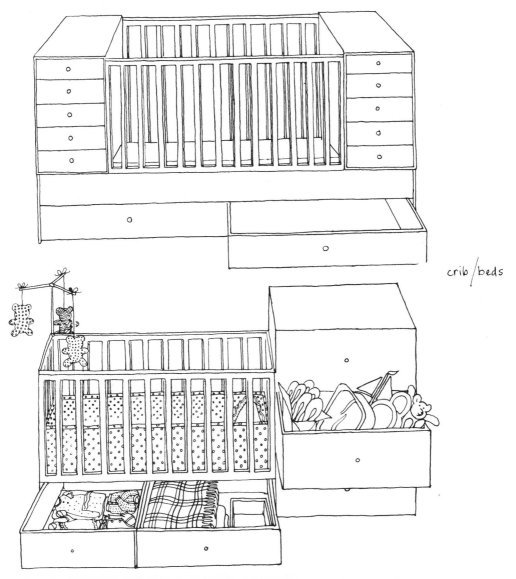

crib/beds

crib/chest combination modular units, separate into a storage bed and chests

trated by a *New York Times* report of an $11 million verdict to the parents of an infant who suffered brain damage when one side of its crib collapsed. Obviously, these guidelines cannot be taken too seriously and, along with common sense, should be applied to all cribs, especially to a borrowed crib or a charming old family heirloom (from your or someone else's family). For example, I recently found two beautiful walnut Victorian folding cribs at a flea market. They cost about the same as new, less interesting commercial cribs, but both had low sides—only about 16" high—which made them perfect for loveseats but not suitable for an infant once he can pull himself up in the crib.

The most important regulation states that crib slats should not be more than 2⅜" apart (roughly the width of three adult fingers), a rule designed to prevent strangulation when a baby wriggles his body through bars, leaving his head behind—or vice versa. If you are using an older crib, such as a charming hand-me-down, check to make sure the slats are close enough together. (Note: The CPSC is very concerned about the safety hazards often presented by older, used cribs.) The crib should be very stable. All screws, bolts, and other hardware must be in place and tightened to ensure the crib is structurally sound and will not collapse or come apart during use. Mattress support hangers should be securely attached to the headboard and baseboard of the crib. Examine it for these other safety features:

All wooden surfaces should be smooth and free of splinters, and all hardware should be free of rough or sharp edges. There should be no projections. On a regular basis, inspect the crib for damage to hardware, loose joints, or missing parts. Do not use the crib if any parts are missing or broken. Do not substitute parts. Use and save manufacturer's instructions on maintenance, cleaning, storage, and use.

Latches on drop sides and folding sides of the crib should hold tightly enough to prevent accidental release. They should require either two actions to open and close, or at least ten pounds of force. There are three kinds of release mechanisms—foot bar, knee push, and double trigger. Sometimes the first two are used together.

The mattresses should fit snugly, leaving no gap between the sides of the crib and the mattress. If you can fit more than the width of two fingers between the mattress and the side of the crib when the mattress is pushed against any one side, the mattress is too small and should be replaced to prevent entrapment of the infant between the mattress and the side of the crib (on the other hand, it shouldn't fit so tightly that you can't change sheets). Most crib manufacturers label a crib with the thickness of the mattress required.

When the mattress support is in its highest position and the crib side is in its lowest position, there should be at least 9" between the mattress support and top of the crib side. This should be at least 26" when the mattress support is in its lowest position and the crib side is in its highest position. When a child reaches a height of 35", the child has outgrown the crib and should be placed in a youth or regular bed.

For protection to the baby's head, and to minimize drafts and keep toys from falling out, use bumper pads around the inside of the crib. Such a pad should be firm, have a washable cover, extend around the entire crib, and tie on in six places, top and bottom. Long, loose ends of ties should be cut short. Secure bumper ties top and bottom and in a way that will prevent a foot or hand being inserted; if a child turns over with arm or leg in the loop, it can act as a tourniquet and be very painful. When a baby begins to use bumpers to climb out of the crib, they should be removed. Ditto large stuffed toys. For infants under 12 months, do not use soft bedding such as quilts, comforters, pillows, sheepskins, or pillow-like bumpers and stuffed toys in the crib. They can present a suffocation hazard.

Note: Mesh, net, and screen cribs are not covered by CSPC regulations. In addition, pre-1960 painted cribs may contain dangerous levels of lead, so older cribs should not be used if possible.

Judging Quality in Cribs

Even among cribs that meet safety standards, there are still qualitative differences. Poor quality control and shoddy construction exist, not just in budget cribs, but in higher-priced "safe" models. So inspect even the finest crib carefully. According to *Consumer Reports* you should look for the following:

Stabilizer bar There should be at least one (two are better) underneath the springs to stabilize the crib against rattling and shaking.

Release mechanism The drop-side release should offer enough resistance so that firm action is required to lower the side. A double release mechanism requiring a foot release as well as an upward tug with hands on the bars is even better.

Teething rails Thin plastic covers that run the length of the top of the railing to prevent baby from chewing on the wood should be firmly attached and not cracked.

Decals Avoid them; they do not stand up well to washing; they chip and peel.

Endboards These may be partially open and ornate or horizontal bars with open spaces that are low enough so a baby's head could get stuck, or easy enough to become a hand- or foothold for a baby who wants to climb out; they are a potential hazard. Straight, functional headboards and footboards are recommended instead—as are endboards without plastic balls (these can be dangerous when broken).

It is always better to buy a crib that the store will install (as many do), and to test working parts such as movable sides on the spot. However, if the crib comes "knocked down" (as many do), check carefully before throwing anything out and before beginning to assemble, to be sure that all parts have been included. As soon as you have done this, remove and discard all plastic wrapping (a potential hazard around young children).

Crib mattresses should have a cover that is durable, that the baby can't rip open, and that is wet-resistant so that it can be wiped or washed off. It should be comfortable—firm yet resilient, not soft and sagging. Since there is little price difference between the best and the worst, it pays to buy the "top of the line," as an inexpensive mattress will not wear well. The two basic kinds of mattresses are coil and foam. Both are equally popular. The key criterion is firmness: very soft sleep surfaces, such as water beds, pillows, and comforters may possibly contribute along with other factors to SIDS (Sudden Infant Death Syndrome) or suffocation. For this reason, doctors advise that babies be placed in cribs on their backs. You can purchase a pillow device for newborns that keeps them there, however the CPSC does not recommend this.

Portable cribs are made from metal, mesh, or wood; they are smaller and weigh less than normal cribs and, as their name indicates, fold so that they can be taken on trips (hence the nickname "Grandma" cribs).

Some are on wheels and narrow, and can be rolled through doorways and used throughout the house as a second crib—along with folding playpen/crib combinations—which is handy in a two-story house, as it saves running up and down stairs.

It is imperative not to leave one side of a mesh crib (or playpen) down, forming a pocket that a young baby might accidentally roll into and suffocate in. Wood portable cribs are subject to similar safety standards as full size cribs, and some models that can collapse and present a strangulation hazard have been recalled.

Adult-Size Beds for Children

When your baby is about to fall or climb out of the crib, it's time for a bed. Waiting too long can be disastrous. "I switched when my child started diving into, rather than climbing out of her crib," warns one parent. Youth beds, slightly smaller than standard beds, are sometimes the next step. However, unless space is limited or you have or are inheriting a youth bed, a child can graduate directly to an adult-size bed with innerspring or foam mattress.

To acclimate a baby to a bed, some parents suggest placing him on it for short periods: for playing, naps, changing; to ease the transition, other parents suggest switching at the time of a trip (a visit to a relative who does not have a crib, for example), or right after you return.

To cushion or prevent harmful falls (the CPSC reported that 13,800 children were treated in 1991 in hospital emergency rooms for falls), consider the following:

Put the mattress directly on the floor surrounded with pillows, or

Lower the bed by removing or cutting off legs, or

Buy a high-riser or trundle bed. Not only can two beds be stored in the space of one but by placing the lower bed next to the baby's bed, he will have a soft place to fall if he rolls off. The adjacent lower bed also provides a step-up into the child's bed and then separates into two beds later for sleep-overs or siblings.

Bunk beds are terrific space-savers and take advantage of unused vertical space. They might be bought ahead if space and pocketbook permit. Two beds, sometimes even three, are placed one on top of another. Many have storage drawers or a trundle, pull-out bed beneath the lower bunk. The advantage of having so much in the space normally occupied by a single bed can, however, be offset by the difficulty in making the beds (unless you are a giant) and the danger of falling off. It is a good idea, if you are buying bunks, to buy a set that comes apart. Then you can use them as separate bed/couches, placing both in the room if there is space, putting the second bed elsewhere in the house if there is not—until it is safe to put one on top of the other. Likewise, if you start out with the upper bunk in place, remove it as soon as the baby starts to crawl and do not put it back until he is an expert climber. (Similarly, an upper-story platform, or loft, can be planned now to go above a standard bed but be executed later on.) A young child, generally under six, should not be allowed to sleep on the top, and there should always be guard rails on all sides. The Consumer Products Safety Commission approved in December 1999 mandatory standards for bunk beds that require guard rails on both sides of any bed, including a continuous rail on the wall side of the bed, with an underside foundation more than 30 inches from the floor. The opening between the rail and the mattress must be less than 3½", and the bed must have a label warning parents not to place children under six years of age on the upper bunk.

Different Things to do With Bunk Beds

When a child is old enough to climb up and down, the upper bunk can be converted to a play area (instead of a sleep area) by eliminating the mattress and adding toys, pillows, a light, and wall storage.

To make a hideaway and "private place," both bunks can be enclosed with curtains or shades hung from within the frame of the bed.

Fabric draped to make a canopy can cover the top of a little girl's bunk bed, until she is old enough to use it for sleeping.

Supplementary Sleeping Space

We've come a long way from the old favorite: the folding cot. Even in a small space it is possible to find a place to tuck away an attractive, permanent, reasonably comfortable bed, useful for a baby-sitter, when a baby is sick or needs extra attention, or for an occasional guest, for a newborn's baby nurse, for a sibling, etc. In addition to bunk beds I suggest:

A twin bed The most convenient size is 39", including a high-riser or a narrower 30"–33" custom-made bed with a removable cover and bolsters. It looks like a couch now, and can become your toddler's bed later on.

A convertible sofa This is less convenient but more elegant, especially if the room is sometimes used as a den; it opens up (allow floor space for this) into a fully "made-up" twin or double bed with innerspring mattress. Some easy chairs and ottomans work the same way.

Beds

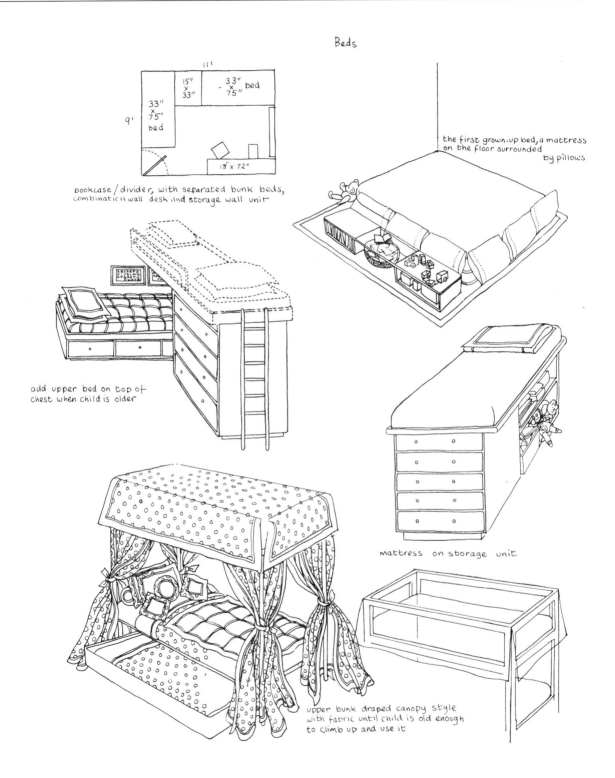

11'

15"
x
33"

33"
x
75"
bed

33"
x
75" bed

9'

18" x 72"

bookcase/divider, with separated bunk beds, combination wall desk and storage wall unit

the first grown-up bed, a mattress on the floor surrounded by pillows

add upper bed on top of chest when child is older

mattress on storage unit

upper bunk draped canopy style with fabric until child is old enough to climb up and use it

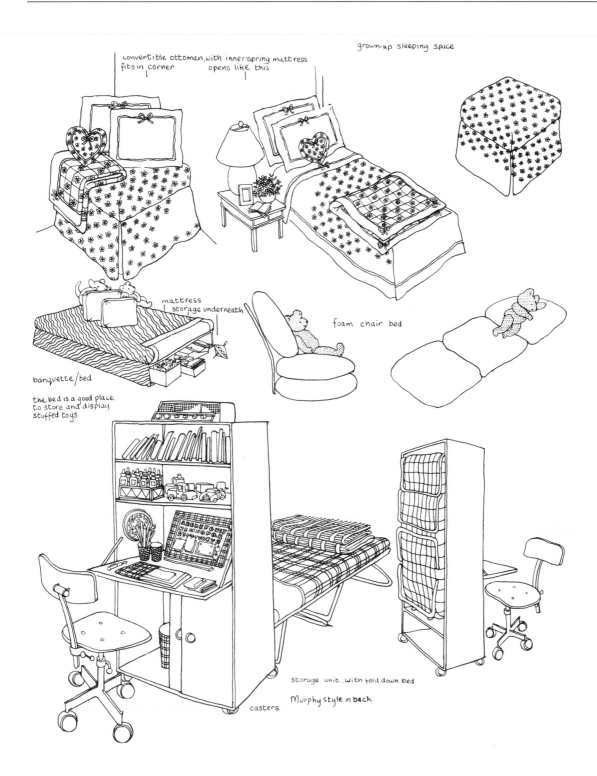

convertible ottoman, with inner-spring mattress
fits in corner   opens like this

grown-up sleeping space

mattress
storage underneath

foam chair bed

banquette/bed

the bed is a good place
to store and display
stuffed toys

storage unit, with fold down bed

casters   Murphy style in back

A sleep sofa or studio bed Consider a foam mattress and bolsters on a base, which must be made up each time. They are therefore somewhat more suitable for naps than for nightly sleeping.

Foam folding couches and futons These are good-looking, moderately priced, lightweight, and safe (no hard edges for a baby to bump into). Foam sleepers convert from couches or chairs when their three sections unfold into a sleeping surface (or a playhouse). A futon, which is a thin cloth mattress, unfolds in a similar way. Both are low on the floor and must be made up.

Murphy, or folding, beds Expensive and somewhat cumbersome, they are nevertheless the *ultimate* space-saver: they flip up and disappear into the wall or a cabinet. If you need floor space rather than space for sitting, these are terrific.

Sleep-over space It isn't necessary to give temporary sleep-over guests too much attention or allocate to them too much space. Kids seem to sleep almost anywhere, and particularly love the floor. So, in addition to the above, select bedding that disappears, folds or rolls up, such as an army cot, sleeping bag, thin mattress, air mattress, or gym mat.

Futon Mattress / Bed
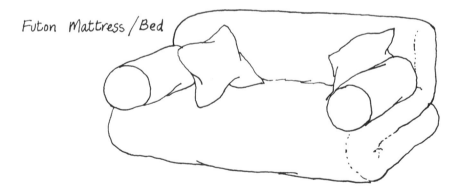

Getting Around: Carriages and Strollers

After a crib, "wheels" are the second most important piece of baby equipment. But which one—carriage? Stroller? Both? Neither? Climate, storage space, and budget are factors to consider.

Think of a carriage not just for airings outdoors, but also for use indoors as a movable bed for an infant or a second bed for naps in another location in the home. In fact, in a cold climate, a carriage can be more useful inside—its sides and roof give protection from drafts when a baby is aired by the window—than outside, when it is too cold for a walk. Since many winter babies spend such a short time in a carriage, it may not pay to buy one. By spring—around age three months, when a baby begins to sit up—he will be ready for a stroller. Parents who live in big cities may find it hard both to store and maneuver a carriage; particularly a large one. Some parents prefer a carriage that is compact, convertible and collapsible, has a body and chassis that folds (some fold into a stroller) and that is removable. And some parents believe in neither a carriage nor a stroller, opting instead for a fabric baby carrier because a carrier requires no storage and allows parents to hold their baby close—either on the back (some backpacks for bigger babies that are sitting have frames) or on the chest (for

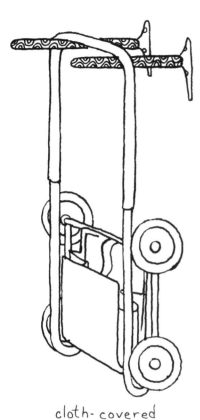

cloth-covered
shelf bracket
to hold carriage

young babies—until the child can make his own way by walking). This is psychologically soothing, and also partially protects city kids from city fumes. These should be chosen according to the size of your child, although some models fit all ages.

Despite the fact that carriages are much more comfortable for a sleeping baby and offer greater protection from the elements, many families with limited space and budget select a stroller as a substitute. A small fold-up, lightweight "umbrella" stroller is adequate, and those that recline can be used from the start. Try to buy one that has a hard back and is not too flimsy and likely to tip. Make sure that, at the very least, brakes are simple to operate and hold securely and that the stroller opens and folds quickly and locks firmly and that it has no potentially harmful protruding parts. A stroller should be easy to push (some have balloon tires and swivel front wheels), and, as is also true for carriages, the handles should be a comfortable height for you.

Do not expect "umbrella" strollers, that are for many babies a second traveling stroller, to hold up as well as larger, sturdier strollers. Better standard models should have these features: tight-locking brakes (two-wheel inside brakes are best), a semireclining seat with adjustable foot and back rest, safety belt, a suspension system for shock absorption, and canopies and sides for weather protection.

If you already have one child, think about how you are going to transport two children when your toddler decides he doesn't want to walk. For this purpose, buy a collapsible seat that fits on top of a carriage (it should be a sturdy one), or a double side-by-side stroller.

Bathing, Dressing, and Eating: Bathtubs, Changing Tables, and High Chairs

Bathing Aids

> For bathing, I suggest you choose one of two alternatives. One is the kitchen sink. This means you will not need to go to any extra expense. You will, however, need to be alert; you must be sure the child neither hits a faucet—thus triggering a spout of possibly scalding water— nor strikes his head (or any other part of his body) on any of the plumbing. A second choice is a plastic tub or dishpan—generally available at low cost— which you can fill at the kitchen sink, then place on a counter or table of convenient height; choose, of course, a size that will fit the child.
>
> VIRGINIA POMERANZ, M.D.,
> *The First Five Years: A Relaxed Approach to Child Care*

Today, according to one manufacturer's survey, 67 percent of all young babies are bathed in the kitchen sink because of the convenience to parents. So it is not surprising to find an abundance of bathing aids that help keep a soapy, squirming, sometimes screaming child "in place." Some fit in a sink or tub, or on a counter. It may be (for up to age six months) an inflatable plastic tub, a small, molded plastic tub, or an actual plastic dishpan. Older babies can be bathed in a larger molded plastic bath placed in the bathtub until the time they are old enough to sit in a baby seat or anchor, or on a molded sponge in the bathtub. A folded diaper or towel placed in the bottom of a sink or tub will prevent slipping.

Changing (or Dressing) Surfaces

The separate changing table, with storage below, seems to be on the same road to extinction as the bassinet. Many parents, whether or not they need to conserve space, seem to prefer changing their child on a multipurpose piece of furniture. Responding to this demand, nursery furniture manufacturers are making a variety of chests with temporary dressing tops (usually part of a set that includes the crib). One style has a removable flip top that folds out for changing. It extends over the chest, making the dressing area larger (these may be easily pulled over, so test this yourself before buying). Another has a removable changing-table top that turns into a stool when no longer needed (see illustration). The waterproof changing pad that comes with it is approximately 32" × 16", has a safety strap (strap or not, Baby should never be left on a changing surface unattended), and can be purchased separately (or homemade, see page 102), and used on furnishings you already own. Any surface that is flat, firm, and high enough (36", or hip height) so you do not have to bend over too far can be pressed into service and either covered with a loose pad or padded with 2" foam rubber cut to size and then covered with a moisture-proof fabric that is nailed or stapled to the under-side of the edges. While chests of drawers and deep shelves are most commonly used for changing, I often hear about babies being dressed on a variety of furnishings never intended for this purpose: utility and rolling carts, architects' tables, antique dry sinks (they already have sides to prevent rolling). As mentioned on page 38, babies can also be changed in a closet if there is space in the lower half to add a deep shelf (see illustration). Not only will you save the space that an

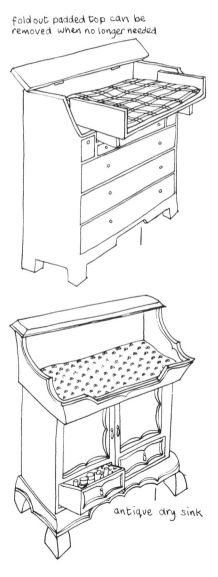

foldout padded top can be removed when no longer needed

antique dry sink

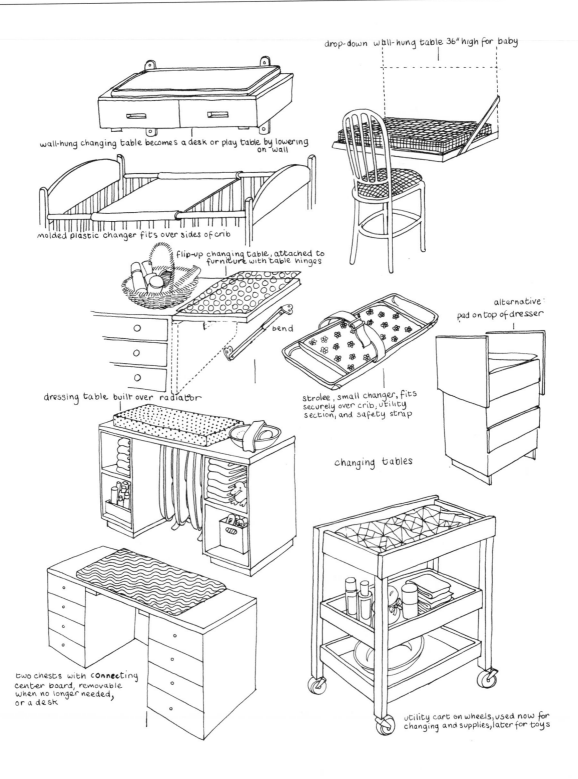

drop-down wall-hung table 36" high for baby

wall-hung changing table becomes a desk or play table by lowering on wall

molded plastic changer fits over sides of crib

flip-up changing table, attached to furniture with table hinges

bend

dressing table built over radiator

strolee, small changer, fits securely over crib; utility section, and safety strap

alternative pad on top of dresser

changing tables

two chests with connecting center board, removable when no longer needed, or a desk

utility cart on wheels, used now for changing and supplies, later for toys

additional piece of furniture takes up (and the cost), but your baby will be safer—less likely to roll off, since the shelf is enclosed on three sides. Similarly, one parent installed a large drop-down shelf over the bathtub; when lowered, it forms an enclosed area for dressing; since it is deeper than her baby is tall, she was able to change her daughter "fore and aft"—Also a deterrent against rolling off.

Another alternative to the free-standing and/or commercial changing table is a wonderful inexpensive molded plastic "changer" that looks like a tray and fits over the sides of the crib. Of course, some parents use the crib or a bed, or even the floor, eliminating the need for a changing table or surface altogether (but adding to your wash load).

High Chairs

Since a high chair is for a child who is sitting up and active, it is not surprising to hear frequent sad stories involving children falling out of them. As a result, consumer regulators insist on certain safety features: stability, a wide, sturdy base, a tray that locks securely, and safety straps that are not attached to the tray (because if the tray were to fall off, the baby would fall with it). A large, high-rimmed tray that contains the tray contents adequately is an advantage, as is a high chair with a tray that is removable so that the chair can be used alone later on.

What old or "antique" high chairs lack in safety—they are not always very stable, do not have restraining straps or eating trays (being simply pushed up to the table)—they more than make up for in beauty. While some manufacturers now make good wooden copies of old pieces, you may prefer to buy a genuine antique high chair. These can be cheaper because they are, at the least, *second*hand. However, be sure to use one that has a sturdy, broad base, add a restraining harness or strap, and keep an eye on your kid when he is in the chair—which is what you should do regardless of style.

Incidentally, the demand for multifunction equipment seems to have spawned a growing group of convertible high chairs that change into youth chairs, play tables, even car and infant seats.

Portable Hook-on High Chairs

Promoted for use when eating out, these seats attach directly to a table or counter and are equally handy at home, particularly if you do not have space for or want a high chair. One caveat: make sure the seat is attached firmly and the table is solidly on the ground and will not tip with the weight of your child. A good alternative, which travels well, is a booster seat that straps to a chair or can sit on the floor.

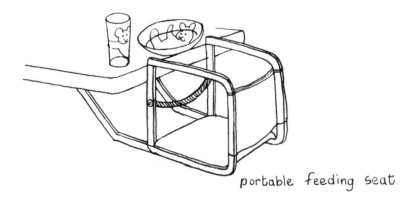

portable feeding seat

Passing Time in Play Yards, Infant Seats, and Swings

Right from the beginning, put your baby into a playpen for some time of the day so that he becomes used to it. If you start using a playpen after baby learns to crawl, it will seem like a jail to it. If you start early, it will be a more acceptable routine. A child should not be left in a playpen for long intervals, but brief stays can be helpful to you when you need to be active and cannot watch him. One great time to make the playpen a habit is when you're cooking dinner.

DR. GIDEON PANTER, *Now That You've Had Your Baby*

Play Yards

Dr. Panter notwithstanding, play yards are very controversial and they have had a name change to prove it. While many people will agree with Dr. Panter, there is a hard core of parents and experts who insist that play yards live up to their former name, play*pen*, by confining children, inhibiting activity and, perhaps, even mental development. They suggest that it is better to let a younger baby roll on a blanket on the floor and an older baby freely explore a house made babyproof (see pages 49–53).

On the plus side, play yards are very practical for people who live in houses—they have more space to use them (including backyards) and to store them. Babies can be kept away from stairs and, in a large home, parents have a convenient place to put a baby near them and their work and away from a rival sibling. (Play yards have even been known to end up as storage yards for toys while the whole house becomes one giant play yard for the child.) Though smaller and even more confining, a portable crib, when it is no longer the principal sleeping space and has its legs removed, can be a play yard.

Precautions for cribs apply to wooden play yards as well: slats should not be more than 2⅜" apart, and you should remove large toys that a baby can use to climb on and out with. Newer mesh and metal play yards or playpens often have extra features, such as a bassinet insert or removable changing table, making them multipurpose.

Infant Seats/Baby Carriers

Useful for a limited but crucial period, an infant seat is a worthwhile purchase. It is usually made of rigid plastic or cloth on a frame, with a stand in the back that allows a baby to sit in a semi-upright position. Infant seats are very convenient for feeding infants who are too small for a high chair and too messy for your lap (some parents say this is its most important use), for carrying and for watching a baby—and for him to watch you. An infant seat should have a restraining belt and the seat-adjustment device at the back should lock firmly in place. However, babies should still never be left unsupervised in them, especially if the seat is on a surface other than the floor that they could fall or slide off, such as a table, chair, couch, or counter.

One model baby seat made of mesh on a springy metal frame has been particularly recommended. The slow motion of a baby's movements sets it gently bouncing.

Car Seats

A car seat is essential for safety if you travel by automobile and is now required by law in most states (including when you leave the hospital). It can probably double as an infant seat but *never* the reverse. A small baby—up to twenty pounds—should be placed in a rear-facing car seat (for newborns, 0–6 months, buy the terry roll that goes inside to make a more secure fit). The baby is held in place with a safety harness and the seat or carrier is securely attached to the car following the manufacturer's instructions. An older baby (1–4 years, 20–40 lbs.) sits facing forward.

Several convertible models become forward-facing when your baby begins to sit up (some even convert to booster seats for the age 4 and older child). These are the most practical to buy since they will last several years instead of several months. Of course sturdiness of construction and adequate padding are of the utmost importance here. A car seat should also recline for napping and sit up high enough so that an older child can see out of the window.

Swings, Bouncers, and Exercise Seats

The 17th century had "rockers," special nurses employed by nobility to rock babies. Today's answer to this is the indoor swing or bouncy seat. They have replaced the now out-of-favor walker, which has been found to be potentially harmful, especially if equipped with wheels. "Swings, either wind-up or battery operated, are not a helpful convenience—they are a *must*," says one mother. "I know several parents who got a peaceful moment during the first six months only because of them."

Today there are a variety of seats that not only provide a place to put baby but also entertain and soothe with sound and movement. "A bouncy seat that vibrates is key for a fussy baby and is safe and easy to transport—I keep one in the kitchen, upstairs, and in the car," said another new mother.

Also available is the Exersaucer or Entertainment Activity Center, which is basically a fixed floor ring with a central seat that allows baby to move around and play with the surrounding toys.

Safety and shopping tips: make sure doorways can safely hold a doorway swing or jumper; consider available space when purchasing an exerciser, which should fold and become portable; never put a bouncer on a high surface; and keep baby seats away from stairs and steps.

All Kinds of Accessories for Infants

It is truly the little things that count, especially in a little nursery—an almost infinite assortment of small accessories can add character while serving a function for you or the baby. Some accessories are absolute necessities: a wastebasket, lamp, and clock; others are conveniences, such as an intercom; and still others are just luxuries that add an attractive touch to the space, such as a colorful picture. Accessories are, however, often the last thing expectant parents think of, and then are frequently bought in a hurried and haphazard way. This is unfortunate, since an accessory can just as easily spoil as enhance the appearance of a room. Here is a list of accessories I recommend, along with some instructions on how to make some of them.

ACCESSORIES CHECKLIST:

A radio, cassette, CD player, or combination for music (digital settings are a nighttime convenience). A clock with large numbers. A shatterproof mirror.

A battery or plug-in intercom, for a newborn and for families that live in a large home

Colorful toys

Wall decorations

Toiletries in attractive holders, baskets, trays

Lamps, a night light, and general lighting on a dimmer
Wastebaskets, diaper pail (a step-on can with a top is good, two at
    least for bedroom and bathroom)
All kinds of decorative nursery coordinates (see pages 107-8)
Nice luxuries:
    A hot plate, bottle warmer, and a small refrigerator, a washing
    machine/dryer combination, a telephone (with the bell turned
    off), a humidifier
Accessories with motion:
    Cuckoo clocks
    Wind chimes hung at window
    Music boxes
    Mobiles
    A bowl of goldfish

Mobiles and Toys

Until a baby is old enough to see beyond to the room around him, the decorations on or near the crib are his only visual stimulus. That is why mobiles are so popular as a baby's first "toy." However, mobiles are not really to be played with or touched. (They should be hung out of reach and removed when the baby can lift himself up and pull the mobile down.) Mobiles are to be watched, and musical ones, to be listened to. They help a baby develop his concentration, and the ability to follow objects as they move through space. Hung securely from or near the crib, or from a hook in the ceiling, a mobile should have fairly large figures, no detachable parts, sharp points, or edges. One currently popular mobile has animals that actually "look down" at the baby. Cut long, loose pieces of string and cords.

Homemade mobiles   Keeping the above in mind, to wires or a hanger (see illustration) tie on different lengths of string or fishing line or heavy thread and then attach small balls, such as colored Ping-Pong balls, strips of colorful cloth, paper or aluminum plates, playing cards or greeting cards, including those you received when the baby was born.

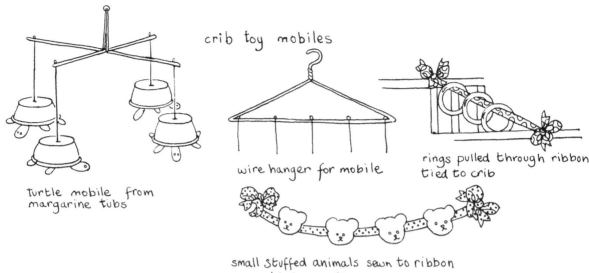

crib toy mobiles

turtle mobile from margarine tubs

wire hanger for mobile

rings pulled through ribbon tied to crib

small stuffed animals sewn to ribbon tied on crib.

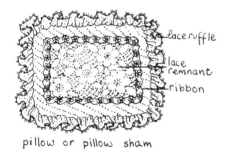

lace ruffle

lace remnant

ribbon

pillow or pillow sham

Crib exerciser and toys For visual stimulation and for touching, pushing, and pulling, simple toys or shapes can be attached to a wooden bar or a strap stretched across the crib. A variation of a store-bought model can be homemade by attaching toys, rings, and similar small objects to a dowel or a piece of elastic. Just be sure the objects used are baby-safe and there are no loose strings or cords that a baby could tangle with.

Stuffed toys These liven up the crib *and* the room (instead of accent pillows, they can decorate a bed). They should be colorful and cuddly, small, and light enough for Baby to handle. Like all toys for small children, they should be sturdy and should not have any parts that can be pulled off and swallowed, such as ribbons, buttons, balls, felt tongues and noses, glass eyes, and whiskers. Simpler stuffed toys with features or details sewed, printed, or painted on are safest and will last the longest.

Homemade tub toys Make them from plastic bottles and empty milk containers. Babies love to pour water in and out of bottles, and the containers float like boats.

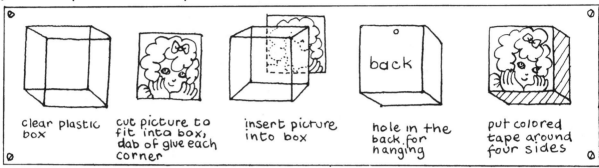

picture frame made from a discarded clear plastic covered box

clear plastic box

cut picture to fit into box, dab of glue each corner

insert picture into box

hole in the back for hanging

put colored tape around four sides

Pictures, Posters, and Other Wall Hangings

Wall decorations can cost very little and take up no space, but they can make or break a child's room (or any room, in fact). On the other hand, "cute"—strictly for the nursery—pictures are commercial and tend to be trite. How much more interesting to hang slightly sophisticated "adult" works of art. Your baby will also surely love:

Colorful posters from tourist and travel offices, theaters, museums, *Sesame Street, National Geographic*s and children's-book publishers;

Framed family photographs or your other children's artwork (a good psychological move);

Pictures from children's books, art books, magazines (and their covers);

Christmas or other greeting cards or postcards;

Animal place mats, kites, flags, quilts, maps;

A boldly designed graphic fabric—or an antique remnant stretched on an artist's stretcher or hung on rods.

Frames and mats can be color- or fabric-coordinated to the room. Colored mats in standard frame sizes can be purchased at many stationery and art supply stores, or a piece of thin fabric, perhaps one used elsewhere in the room, can be glued onto an existing mat. Colored tape or paint can cover an old frame. Pictures look best if they are hung in a group rather

plastic place mat on a wall

than separately scattered around the room. To hang pictures, use small nails or picture hooks on walls. To picture frames attach gummed-paper hangers, screw eyes and wire, small metal hangers (nail), a piece of wire, cord, or yarn (staple them or thread them through screw eyes—the choice depends on weight). Pictures without frames can be hung using double-faced foam tape or tabs (in effect, a way of "gluing" the picture to the wall) or small clips with holes.

A blackboard may now be a "white" magnetic drawing board with a porcelain enamel surface, available in many sizes (buy a large one, 3' × 4' for the nursery). Until your baby can contribute his own artwork, you can draw or write on them with colorful washable and erasable felt-tip markers, or attach messages, such as important telephone numbers for the baby-sitter, with magnets.

hang a kite from the ceiling.

bulletin boards, pictures, and wall hangings

brackets or nails
cardboard covered
with fabric.

ribbon trims an
old bulletin board

antique frame

mirror or bulletin board

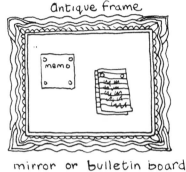

paint
frame

wide ribbon attached
to wall,
pictures hang over it

Bulletin Boards

These can carry through the decorating scheme. Store-bought boards can be painted all over or only on the frame. Make your own with a sheet of Homosote or two pieces of corrugated cardboard covered with the same fabric as the nursery coordinates or a piece of bed sheet, or wallpaper. Trim with ribbon or colored tape and add colorful tacks (keep out of a baby's reach). Since your child's future artwork will surely be oversized, the bulletin board should be big—at least 24" × 30", or larger. Place a board between two doors or two windows, or cover an entire wall.

Flowers and Containers

Babies love the color and slow "motion" of flowers, and even a single bud placed near enough to watch—but not eat—will fascinate them.

Put flowers in a pretty bottle left as is, or spray-painted, or in baskets and boxes (place flowers in a glass first, then place the glass in the basket or box).

Flower pots can be decorated with paint, fabric, wallpaper, stickers, and other small decorations.

Fixing Up Baby Space

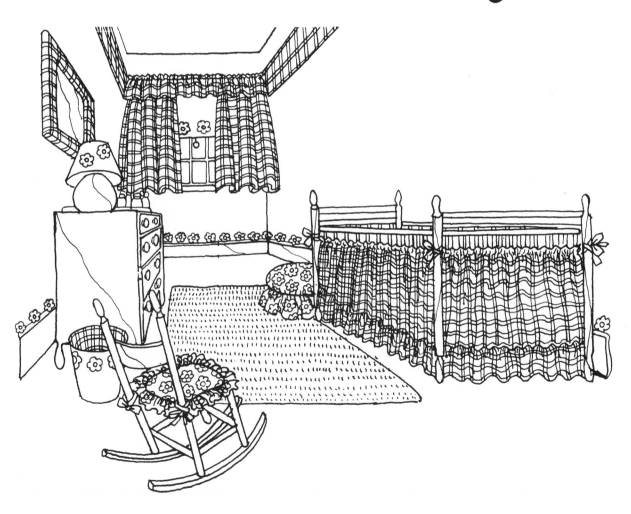

Introduction

Nursery design is summed up in three words: simplicity, passion, and color. And there should always be elements of humor, surprise, and comfort.

PEARL THOMPSON, Designer
Nursery Lines, LTD, New York City

Even though the decoration is for the parents' pleasure, a child will appreciate and be affected by a clean, colorful, and congenial environment. The benefits, however, have to be carefully measured against the price. In the euphoria often attendant upon the arrival of a first child, parents may forget that it costs as much to finish a floor or hang wall covering in an infant's room as in their own. It's not hard to end up investing more than you ever would have dreamed, only to find that your child has quickly outgrown the decoration and you have to redecorate.

This section discusses each of the elements that compose the decorating scheme for baby space, from the choice of a color scheme to the covering of the ceiling. There are many inexpensive ways to achieve the same expensive effect a decorator achieves, especially if you do it yourself. So I have included instructions that can be implemented by you with such familiar modern marvels as a staple gun, white glue, no-sew, iron-on bonding tape and Magic Markers, with or without professional help (which is addressed in the last chapter).

These instructions are oversimplified for parents who are not perfectionists, who want not so much full blueprints as inspiration and ideas. (More detailed instructions are always available in how-to books and from the salespeople where you buy your supplies.) In several places alternative instructions are suggested: a way to make a round cloth by carefully measuring and sewing, or a way to make it by casually throwing it over a table and then cutting to size; a way to hang curtains on rods or by simply stapling them onto the window frame.

Color, Fabric, Pattern

The girl that I marry will have to be
As soft and as pink as a nursery.

IRVING BERLIN

It's no longer simply pink or blue. Choosing a color for the nursery is more controversial and complicated now that feminists and scientists have entered the room.

Psychologists generally agree that certain colors perceived as beautiful—light blue, yellow, light green, orange—have a decisive positive influence on a child's IQ, make him friendly, alert, creative, while those perceived as ugly—white, black, brown—have a negative influence. And even though he is color-blind until about the fourth month, the experts say a newborn will be most secure in softly colored pastel surroundings. They can, however, distinguish brightness, they do respond well to strong contrasts and bright colors, such as primary colors (as do most toddlers). Cool colors, such as blues and greens, are supposed to be more relaxing than hot colors such as red and orange that are more stimulating. So psychologists recommend putting a passive child in a room with a lively, bright color scheme and a hyperactive child in a room with a "peaceful" color scheme. This was borne out in an experiment once conducted in a Canadian school by Harold Wohlfarth, a photo-biologist and president of the German Academy of Color Science. He found that when the schoolroom colors were changed from orange and white to blues and grays, the children were less aggressive and much better behaved.

Of course, you can sidestep the color question completely by opting for all-white or all natural colors, varying texture and adding colorful accessories.

If you find the above a bit confusing and somewhat obvious, now add to it the more subtle sexist dilemma. In *Growing Up Free,* Letty Cottin Pogrebin advocates not just a nursery but a whole home free of gender color-coding.

When colors are linked to gender, the rooms of a house become ghettos—as divided and alienated from each other as the urban ghettos that separate blacks and whites.

This is not only divisive, it is downright uneconomical. With wallpaper, bedspreads, curtains, carpeting, and furnishings, just as with clothes hand-me-downs do double duty if they can be passed from girl to boy or vice versa without a dye job or a paint job. . . . My only concern is to make you aware of the areas where sex-typing can become an unnecessary impediment to your efficiency. Instead of looking at practicality and durability, many people focus on incidentals: color and decoration.

What if the best buy in a crib bumper happens to be available only in blue? Suppose your heirloom bassinette is pink-painted wicker laced with pink satin ribbons? Do you tell your parents to keep it in the attic? Do you repaint it?

Your answers will depend on whether you choose to revise reality, or to neutralize it. . . . The Revisionist Position: First, you resolve to be color blind. Then you buy or accept any item you like, regardless of its color, drawings, or ornamentation . . . You don't care if the lamp in your baby daughter's room is mounted on a statue of a Space Man, so long as the lamp is cheerful and provides light. . . .

The Neutrality Position: Very simply, ban pink and blue from babyland. Insist on such gender-neutral infancy pastels as pale yel-

low, lime green, or white, or search for furnishings and baby clothes in red, royal blue, kelly green, and bright yellow—the colors that babies find most appealing and adults find least sex-linked.

As for decoration and ornamentation, the idea is to provide all sorts of visual stimuli so that, as children develop consciousness, many you-can-do-anything images are within view: airplanes and dollies on the playpen mat, for instance . . . On the subject of pink (which today can also be mauve, violet, orchid, peach, and blue or gray or tan or brown) . . .

Girls are not meant to be bright—or vigorous or black, brown, or yellow. Girls are meant to be pink. The "feminine" ideal in the language of the palette. Petal pink and delicate. Cotton-candy pink and sweet. Baby pink and tender, vulnerable, fragile, dainty, and helpless. The male version of that sappy message is, of course, pink stinks. . . . Actually, what I hate is what happens to pink, when it's called into service to decorate a little girl's room or clothing. It colors ruffles that cannot be sat on, and lace that costs a fortune to clean. It also seems to coordinate with spindly-legged furniture and fabric that tears easily, thereby restricting girls' mobility and activity as effectively as shackles and chains. Pink is warm. But pink is suggestive of delicacy, of small, controlled motions in a small, confined space.

Blue is cold. Blue is the color guard for no-nonsense "masculinity." A "practical" blue decor can make a four-year-old's bedroom as inviting as a U.S. Navy recruitment office. Or blue can be the wild blue yonder; the sky behind the airplane print curtains or the ocean on which clipper ships sail across the wallpaper. Blue suggests that the big world out there is relevant to the little boy growing up in that room. Blue hides a multitude of stains.

After taking all this psychological information into account, you will probably do as I would and choose a color(s) that appeals to you aesthetically. It may even be a soft blue, which is now very popular for either gender. The basic color of the nursery, and the color scheme that will be built around it, may be inspired—and usually is—by something already in the room, such as a rug or picture, or by a pretty piece of wallpaper or fabric. For example, if the curtains and crib dust-ruffle fabric are yellow, green, and white, you can have yellow walls, a white-painted floor (keeping the larger background areas light will make the room look larger) with green accents appearing in the throw rug, seat cushion of rocker, and pillows on the daybed. The chart below illustrates the proportionate amount that each item in the color scheme requires. Use this by tracing, then filling in different schemes you are considering; or as a guide, by placing actual swatches in corresponding sizes next to each other and then stapling them to a piece of cardboard. Another way to preview your color scheme would be to roughly trace one of the room settings in this book that most approximates your nursery and then fill in colors with thin colored markers or crayons.

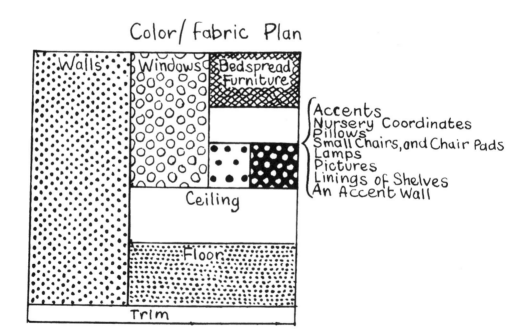

Color/Fabric Plan

Fabric and Pattern

The custom of draping fabric on everything in sight—windows, walls, doors, furniture, even fireplaces—dates back to the 19th century. And it's a tradition worthy of being followed: lots of fabric and only a little time and effort (or none at all if you buy ready-mades or have things custom-made) can almost instantly make a big impact, even on a small budget, and especially in a small space. For children's rooms the best choices are: washable cotton or cotton/synthetic blends, along with inexpensive dress fabrics, canvas, denim, corduroy, ticking, chintz, terry cloth, Haitian or Indian cotton, calicoes, Paisleys, ginghams, country floral prints, lace, and muslin. Old lace linens such as bedspreads, napkins, tablecloths, runners, pillowcases, antimacassars and lace pieces from clothing such as shawls, collars, and cuffs can be recycled, carefully hand-washed and bleached if necessary; reinforced and patched with newer pieces of lace or trim; or used "as is." Washable trims such as eyelet, rickrack, ribbon, and new lace add a nice finishing touch to fabric furnishings. If you must use a nonwashable fabric or have an upholstered piece covered with nonwashable fabric, protect it with a stain-repellent finish.

Make the fabric from coordinated nursery ensembles the basis for a knock-out nursery. An instant unified effect is easy since all of the following are made to match: crib skirts, bumpers and bumpers with padded headboards, sheets (tops and fitted bottoms), pillows (not safe in a crib, nice on a bed or chair), quilts, blankets, comforters, high-chair and infant-seat covers, canopies, toy or diaper holders, lamps. Now even wallpaper and extra fabric match, to cover everything else that's left uncovered: furniture, chair seats, wastebaskets, etc. Of course, all of the above can be made to order (you give the store your fabric and they make it), or homemade (see also chapters on windows, wall, furniture).

Homemade Fabric Furnishings

Crib skirt Roughly measure fabric to fit over the mattress support (underneath the mattress) and down to the floor. Adjust the length, trim, and hem, if necessary. Use extra sets of curtains as a crib skirt.

Bassinet skirt Applying the same principle, you can make a skirt for a bassinet. Push the mattress down over fabric to hold it in place; if necessary to secure fabric, tack it to the basket with strong thread and a long needle. You can also tie a long narrow length of fabric around the basket, or thread narrow pieces of fabric or ribbon through the basket.

Bumpers Estimate the amount of polyester fiber filling or Dacron padding and fabric needed by measuring perimeters of crib all around. Multiply this by 6" plus seams for width. Add six ties, or attach to crib with velcro.

Crib cover or sheet Use actual sheets or something similar. For turn-under, cut fabric 3" more than mattress on all four sides. Sew elastic around fabric like a shower cap, or sew corners together.

Changing-pad cover Make a removable cover like a pillowcase. After pad is inserted, fold over and tack or close with snaps. Use a vinylized or waterproof fabric; or use any washable fabric, or a pretty pillowcase, and place a protective small rubber sheet on top.

No-sew throw or floor pillows Starting with an old pillow, cut a length of fabric about three times its size. Lay fabric wrong side up and place pillow diagonally in the center. Fold one corner of the fabric over the pillow, turning raw edge under, then fold the opposite corner in the same way. Tie the two remaining ends together in a square knot (right over left, left over right). Tuck in ends beneath the folds of the fabric and flatten the knot. *Note:* Instead of fabric, you can use cotton scarves for small pillows (see illustration).

Drape-and-tie instant upholstery To make a throw that "ties on" chairs and couches: tape-measure furniture to be covered, front to back and side to side from the floor. Seam together enough lengths of fabric to make a "dropcloth" 24"

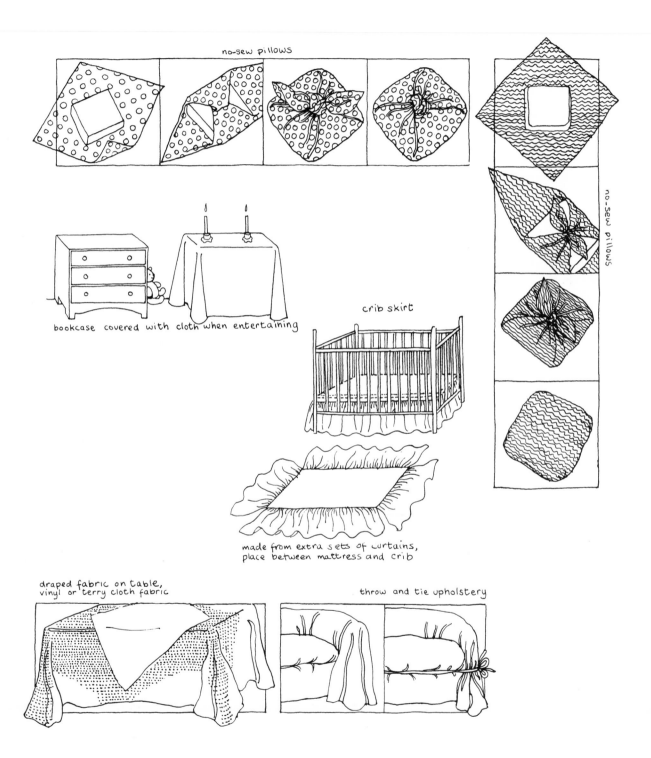

no-sew pillows

no-sew pillows

bookcase covered with cloth when entertaining

crib skirt

made from extra sets of curtains, place between mattress and crib

draped fabric on table, vinyl or terry cloth fabric

throw and tie upholstery

longer and wider than your measurements. Drape on chair, tuck in deeply around seat. Smooth, gather excess at corners and pin. Trim bottom evenly, allowing for hem. Mark center, front and back on underside. Remove, hem, and replace. Gather fabric at corners, and sew lightly, attach bows.

A special shortcut for those in a hurry who hate to measure is to make this and other throws by literally throwing or draping fabric over bed, couch, and table, pinning on extra panels of fabric for more width, if necessary. Cut around the bottom (or just tuck extra material under), pin up, remove, hem (use no-sew iron-on bonding tape), trim, and replace (see illustration).

Round table cover This can conceal an ugly old table. A round table can be made from a large spool for cable or a sturdy stool.

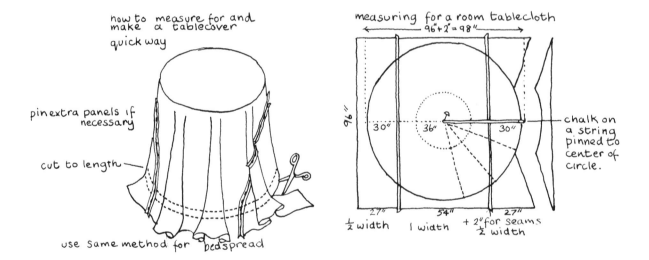

how to measure for and make a tablecover

quick way

pin extra panels if necessary

cut to length

use same method for bedspread

measuring for a room tablecloth

96" + 2" = 98"

96"

30"   36"   30"

chalk on a string pinned to center of circle.

27"   54"   27"
½ width   1 width   + 2" for seams
                        ½ width

To calculate the width and length (this of course is the same in a circle) of fabric needed, measure the table from the floor up, across the top of the table, and down the other side and add extra inches for seams or hem, if necessary. Sew together more than one width of fabric, adding full or half widths to each side, to make a piece wide enough to cover the table down to the floor. Or to avoid piecing, use extra-wide fabric such as felt or a large queen- or king-sized sheet. To make a giant compass, tack one end of a string to the approximate center of the fabric (fold fabric in quarters to find it), and chalk or crayon to the other end. (The length of the string will be the radius of the circle.) Draw circle and cut. Glue or sew on a decorative trim to cover a raveled edge. (Felt will not have to be fringed because it does not ravel.) To cut down laundering the large cloth, cover it with a smaller square of washable fabric, or a round piece of glass. Similarly, make a cloth cover for a square table or a table or a table-height toy storage unit that you wish to camouflage.

Bedspreads   Make a simple floor-length throw by measuring the length from the foot of the bed to the head and from side to side from the floor. Add about 15" for pillow tuck-in and 4" for hems. To avoid a seam down the center, piece at the sides, or use extra-wide fabric such as a king-size sheet. Spreads for bunks are even easier. Use fabric wide and long enough to tuck in. (Hems may not be necessary since fabric is tucked in, unless it ravels.) Trim with fringe, ribbon, or blanket binding. A warm fabric bedspread (such as wool) can double as a blanket.

Slip seat   For a rocker or occasional chair. Unscrew seat, cut fabric, using the seat as a template and allowing two inches for turn-over, staple on fabric, stretching it tight. Screw seat back on.

Canopies over a crib   (1) Hang a width of fabric on rods hung trapeze-style from the ceiling (see illustration). (2) Attach a pair of floor-to-ceiling-length curtains (or sheets) to a curtain rod installed near the ceiling above the crib. Tie back with ribbons that are nailed or stapled to the wall (see illustration). (3) Run long lengths of fabric through five large rings attached to the ceiling. Swags can also be stapled up (see illustration). *Note:* Hardware for canopies should be firmly secured to the ceiling and a canopy should be removed before a baby is old enough to pull it and remove it himself.

Special note: Buy Simplicity Pattern #125 for a set of patterns and instructions on how to make *Ten Baby Basics*, including

Canopies
short cut: can also
be stapled to ceiling

dowels or rods

Sheila Camera

hooks in ceiling

string

dowels or rods

five rings screwed to ceiling

two lengths of fabric

bumper pads and diaper holder, and #126 for instructions on making nine canopies and a bassinet skirt and hood.

Fabric Accessories

Bits of extra fabric can cover hangers, fill the inside of an odds-and-ends box or basket, be used to make a laundry bag, diaper stacker, or a pin cushion (see illustration).

With diluted white glue and spray adhesive, fabric can be attached to and wrapped around the following: lamp shades, boxes or trays to hold toys and toiletries, picture frames, wastebaskets. On large areas such as chests or shelves, spray glue on, or spread glue with a small paint roller, then smooth over fabric with a rolling pin or a large soda bottle.

fabric sleeve for wire hanger

Make Your Own Fabric

Decorate plain fabric with permanent Magic Markers, fabric color sticks, or fabric paint (available at art stores), or crayons (to make permanent, iron crayon side down onto newspaper), or stamp it, substituting acrylic paint for a stamp pad (use a store-bought stamp or let a kindergarten-aged child show you how to make a stamp from an empty spool or a cut-up potato). Liquid dye can also be painted on freehand or used with stencil and stencil brush.

Revitalize old fabric and fabric furnishings—curtains, rugs, bedspreads—with new trim and by dipping in a dye bath (color of dye should be darker than the color of material to be dyed).

Change the color of old welting and trim with permanent-colored markers. Use them also to fill in color in upholstered furniture (or low-pile rugs) where threads are worn or discolored.

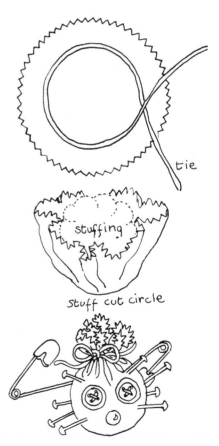

tie

stuffing

stuff cut circle

pin cushion

How to Mix and Match Patterns

Repeat a larger-scale pattern in the room, e.g., squares or circles, with a small-scale pattern of the same thing.

Use the same pattern in two different colors, or reverse: red dots on white background, white dots on red background.

Put together similar patterns in the same color, such as small geometries or florals.

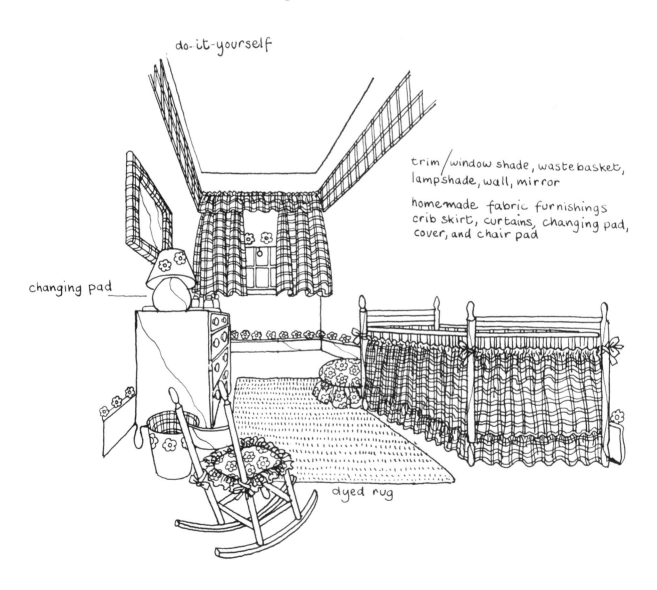

do-it-yourself

trim/window shade, wastebasket, lampshade, wall, mirror

homemade fabric furnishings crib skirt, curtains, changing pad, cover, and chair pad

changing pad

dyed rug

Rejuvenating Old Furniture

The aesthetic sensibilities wake early in some children, and these, if able to analyze their emotions, could testify to what suffering they have been subjected by the habit of sending to schoolroom and nurseries whatever furniture is too ugly or threadbare to be used in any other part of the house.

EDITH WHARTON,
The Decoration of Houses (1897)

Many babies today also inherit cast-off furniture for their first room. However, it's not all that bad. Hand-me-down furniture is often sturdier than a new piece and offers a creative parent many decorating opportunities. Of course the following ideas can also be applied to new unpainted, even finished furniture.

Modernize old chests, cabinets, and chairs with new moldings and new hardware, such as brightly colored drawer pulls.

Completely cover with paper (vinyl paper will make furniture washable) or fabric, using diluted white glue or spray adhesive. Or just cover the top and drawer fronts.

Decorate with designs cut from fabric, wallpaper, magazines, or with stickers, colored tape, or Magic Markers.

Stencil furniture, paint a design freehand, paint stripes (using masking tape to make the stripes), paint drawers different colors, or paint a picture of contents, e.g., socks, shirts.

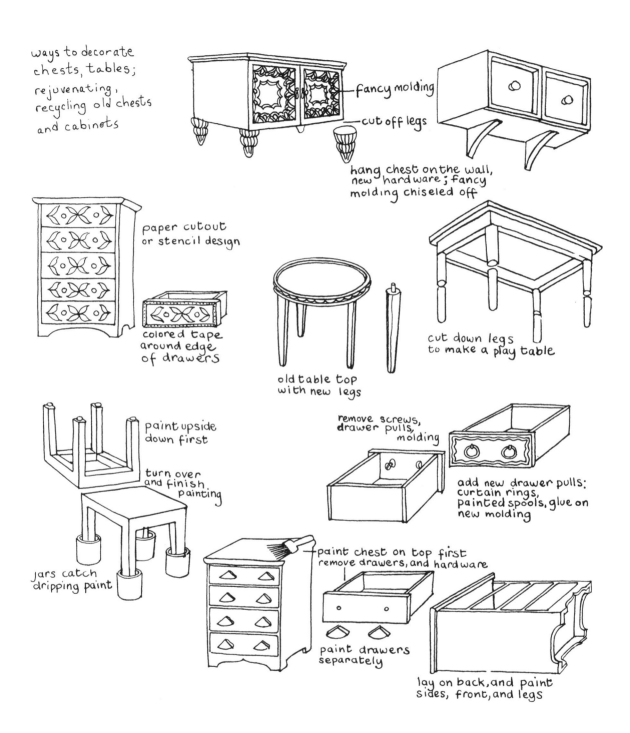

ways to decorate
chests, tables;
rejuvenating,
recycling old chests
and cabinets

fancy molding

cut off legs

hang chest on the wall,
new hardware; fancy
molding chiseled off

paper cutout
or stencil design

colored tape
around edge
of drawers

old table top
with new legs

cut down legs
to make a play table

paint upside
down first

turn over
and finish
painting

jars catch
dripping paint

remove screws,
drawer pulls,
molding

add new drawer pulls:
curtain rings,
painted spools, glue on
new molding

paint chest on top first
remove drawers, and hardware

paint drawers
separately

lay on back, and paint
sides, front, and legs

decorating furniture

needlepoint stencil, sew on appliqué, or color with markers

leftover floor tiles for table top.

chest and mirror, covered with fabric, painted, or papered to match curtains, cut out shapes from fabric that is used on bed or for curtains, and glue them all over

Other painted finishes include antiqued, splattered, and mottled. A high-gloss enamel paint finish is very glossy and durable; an automobile paint is even more so (yes, as on your car, it must be applied with special equipment that blows the paint on).

A plastic "paper" or a glossy paint already has built-in protection; however, a coat of polyurethane plastic should be sprayed or brushed over other paper or fabric decorations for longer life.

Restoration Tips

1. Before refinishing, reglue and brace rickety pieces (remove old glue before regluing, if possible) and fill in dents or holes with wood filler.

2. Smooth filled-in holes, bumpy or peeling areas, and rough surfaces with sandpaper. A shiny surface that is to be painted should also be roughed up a bit with sandpaper to make the paint adhere better (fine sanding is also recommended between coats of glossy paint).

3. All furniture being finished should be dry, smooth, and clean, free of wax, grease, and dirt.

4. Sometimes a good wiping with a clean cloth and mild soap, a quick retouching of scratches with crayon, eyebrow pencil, shoe polish, or scratch cover, and a thorough waxing or polishing are all that is needed to fix up an old finish.

5. Paint a chest or cabinet in this order: remove hardware and drawers and paint drawers separately; paint chest top; then lay chest on its back and paint sides, front, legs.

6. Don't forget the inside of doors, inside and outside of drawers, which can also be covered with wallpaper or fabric to match the room. For odd shapes, such as the inside of a drawer or breakfront, make a paper pattern first, then cut fabric or paper with pattern as guide.

Windows

Even if nothing else is "decorated" or changed, just a window treatment alone can give a room character, establish a theme, and enhance even the smallest of spaces. A window treatment can be frilly and soft—a curtain with ruffles; or tailored and crisp—a window shade, horizontal or vertical blind or shutters; or a combination of both. The window material should be washable, or at least wipeable; it may be transparent (light and vision going through), translucent (only light going through), or opaque (neither going through). The amount of light should be easy to adjust. Deciding on how dark the room will be is up to you, as an infant can adapt to almost anything; however, an older baby may need total darkness to take a nap.

frill tiebacks

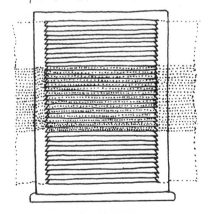

three colored horizontal blind, paint wall to match

Curtains

(See also "Color, Fabric, Pattern," pages 97–108.) Heavy, floor-length draperies might be energy-efficient and keep the nursery toasty warm, but they will also "close it in" and take up space; instead use lightweight simple curtains (either sill or full length, but never in between) to give the nursery a feeling of spaciousness.

Standard sizes for curtains are:

Length for cafes: 25", 30", 36", 40".

Length for other curtains and draperies: 45", 54", 63", 72", 81", 90". One standard drapery width if 48", pleated measurement.

Since skimpy curtains look cheap, make sure you use at least twice the width of the window for normal fabric and three times the width if the fabric is sheer or thin.

Several different traditional window treatments are illustrated, along with some unusual ways to dress a window described below. These are casual, easy-to-make, and perfect for temporary rooms.

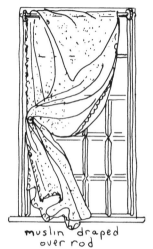

muslin draped over rod

A panel of fabric the same size as the window frame (with allowances for hem) can be hung flat (this takes less fabric than gathered curtains). Hem all around (not necessary if fabric does not ravel or has finished edges). Hang with curtain rings, or pretty cord or ribbon sewn on, then tied in bows to rod; or attach with grommets (on fabric) and cup hooks (on window frame), or just staple to the window frame. For air and light, pull a corner up, attaching it to a hook or something similar installed in an upper corner of the window frame (see illustration). Since the underside will show when the panel is lifted up, it might be nice to line the panel with a contrasting color.

Or a panel of filmy fabric such as lace or muslin, about one and a half times the length of the window and as wide, can be gracefully draped over a curtain rod (see illustration).

For an instant installation that requires no sewing or nailing, use iron-on bonding tape to make hems, no-sew/clip-on curtain rings, and a spring curtain rod (adjust the length and snap it into the window frame).

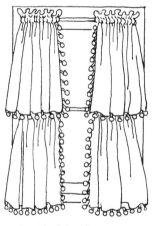

ball fringing

Simple curtains can be made from lengths of fabric in the desired number of widths, depending upon whether you want a full curtain (two or three times the window frame), or flat panel treatment. Hang by simply hemming fabric top and bottom (hems and a channel for rod may already be in sheeting) and shirring it on a rod.

Customize even the cheapest plain-jane pair of curtains by adding fringe or trim. A too short standard-size curtain can

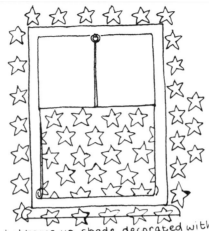

bottoms-up shade decorated with fabric cutouts

tie back with trim or ribbon

flat panels of fabric

held by hooks, and grommets. Panel has contrasting lining

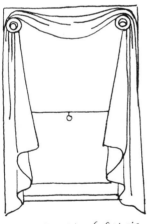

swag a length of fabric

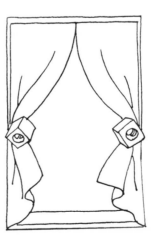

tie-back with toy block

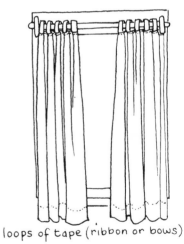

loops of tape (ribbon or bows)

be lengthened with a wide fringe. Other ways to adapt standard sizes to your odd-size window include moving the rod, shortening by hemming, and lengthening by adding a valance.

Roller Shades

Easy and fun to decorate and/or make, window shades generally cost less, take up less space, material, and light (they roll up and "disappear" whereas curtains, particularly tiebacks, block some light). Shades are normally attached to the top of a window, either inside the window frame or on it (more air- and light-tight); however, they can also be installed "bottoms up"—rolled up from the bottom of the frame, thereby giving light above and privacy below. Shades are available in a variety of colors and materials, usually cloth or vinyl, but also bamboo and wood (matchstick) or material that simulates them.

New or old shades can be decorated with cutouts from fabric and wallpaper, with a design made with permanent Magic Markers or painted. (Do not cover the entire surface of a shade with paint, because it has a tendency to crack with movement.)

Shades, like curtains, can be trimmed at the bottom, and the same trim can be glued around the window frame.

Some stores sell shade kits that consist of a wooden roller, brackets, a bottom lath, cord, tacks, and glue. While it helps to have a kit that includes all necessary ingredients, it is not hard to make a roller shade without one, *almost* from scratch. A thin, tightly woven fabric (best because it will not stretch and will hang better) cut to shade size can be glued with spray adhesive or diluted white glue onto an existing cloth shade. A heavy fabric or cloth-backed wallpaper or self-adhesive plastic (with the backing left on), can be stapled or tacked directly to the roller without a shade backing, as can thin fabric sprayed with a stiffener (also cuts down fraying) or laminated onto a special shade cloth (by you or professionally). Of course, the entire shade can be custom-made, using your fabric. Finally, don't forget to make shade and fabric for shade about 12" longer than the window to allow for "roll-up."

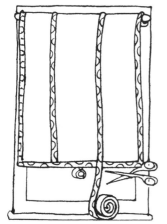

decorate a shade with colored tape

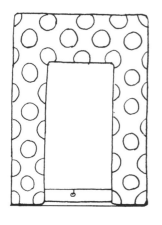

plywood piece cut out and nailed on frame

Blinds and Shutters

Blinds are sleek and contemporary, easy to clean (that's good, because horizontal blinds do collect dust), control light most effectively and can be made to order in the colors of the room. Vertical blinds are wider, take up a little more room and are usually installed floor to ceiling—concealing windows, air conditioners, radiators—over a span that could be the width of an entire wall.

Mini slatted vinyl or aluminum horizontal blinds are very popular and reasonably priced for the amount of service and decorative dash they deliver. Since the slats are narrow, these blinds fit neatly on or in window frames (although they can cover as much wall as necessary), and are available in an extraordinary range of colors and in a stripe of several colors. One caveat: to avoid a toddler's tangling with them, tie up long cords or hang them high on a hook or cleat.

Shutters are charming but they are not cheap, unless you have them already. In that case paint them to go with the room or stain them to go with the finish of the furniture. Part of the cost of shutters is labor, as it is hard to get a good fit within a window frame. One way around this is to make a new, separate frame the size of the shutter and attach the entire unit, frame and shutter, to the window frame.

Easy Ideas for Windows

Hang panels of new or old lace straight down or tied back.

Hang bells, chimes, or mobiles; babies love to watch and listen to them.

Stencil the same border around windows and on window shades or curtain tiebacks.

Weave ribbon or cord in room colors through matchstick blinds.

String colored beads or spools onto cord; tie cord onto cup hooks or screw eyes in the top of the window frame and hang (these also can be a room divider).

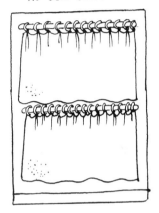

no sew towels

no-sew sheet

cut open hem of sheet to make channel for rod

Make no-sew curtains from matching towels, sheets, or pillowcases; hang by putting a rod through an existing hem or with clip-on curtain rings.

Tie curtains back with string, fabric, or felt, used singly or with two pieces twisted or three pieces braided.

"Write" child's name on a window shade with colored tape (notch tape to make it curve).

Attach ribbon, ties, or tape to a panel of fabric.

Appliqué a motif cut from a lightweight patterned fabric; glue onto shade with spray adhesive or diluted white glue.

Make a swag from a long length of fabric roughly two window heights, plus one width, plus one yard; rings or rosettes attached at upper corners of window will hold the fabric back.

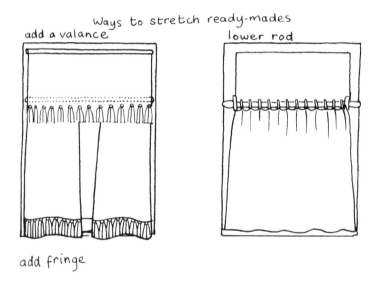

Ways to stretch ready-mades

add a valance

lower rod

add fringe

Lighting

We all know that children need good lighting for healthy eyes; but we don't always achieve it in their rooms. There should be a combination of general lighting—a wall fixture or an electrified track with attached light fixtures, or else a big overhead fixture; and task lighting—a table, desk, or wall lamp for reading and rocking. Some lighting, even a reading lamp, should be controlled by a switch at the door on a dimmer and low enough for small children to reach. A night light, or a fixture with a very low-wattage bulb, is a convenience for you now in the nursery and will give a sense of security to a growing child in the future. Since both table and floor lamps can be easily knocked over by a child, I recommend using lamps that clip onto shelves or wall lights instead (just don't put one too near the crib, in arm's reach of a baby who could burn himself on the bare bulb, or worse). Also, wall lights take up essentially no room. Those that have flexible arms, such as Luxo lamps, can swing around to light different areas (for example, both the desk and the bed next to it), thereby cutting down on the number of lamps needed. However, you may have to hold off buying metal fixtures like these that can become too hot to touch until Baby is old enough to know better. Your baby will also appreciate it if you check that there is no glare or light shining in his eyes from bulbs that may be protected from your view but not his angle of vision.

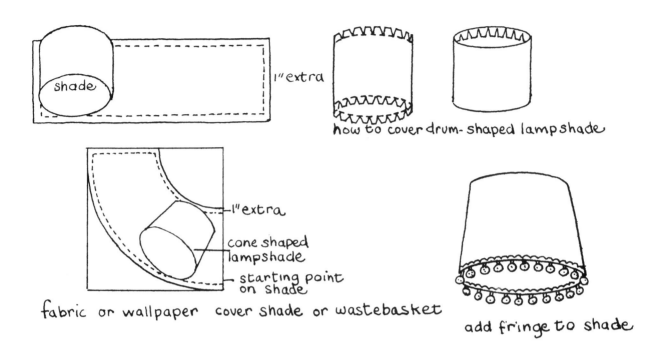

shade

1" extra

how to cover drum-shaped lampshade

1" extra

cone shaped lampshade

starting point on shade

fabric or wallpaper   cover shade or wastebasket

add fringe to shade

Fixing Up Light Fixtures

Old lamp bases and existing wall fixtures and light switches can be renewed and decorated with spray or brush paint; trimmed with small toys, tape, ribbon, stickers, fringe, and so forth.

Old or new lampshades can be covered with fabric or wallpaper as follows (see illustrations): to make a pattern of the shade, remove old fabric from a lampshade. Use the piece that you remove as your pattern or make a pattern by wrapping shade in old newspaper or wrapping paper and then cutting to size. Pin pattern on fabric (or outline pattern with a crayon) and cut. To cover without a pattern: if your lampshade is in the shape of a drum, cut covering material 1" wider and 1" longer at top and bottom for overlap. If your lampshade is in the shape of a cone (larger at bottom than at top), mark starting point on shade, then roll the shade on the covering. Then mark shape as shown and cut out (see illustration).

simple ceiling fixtures

do-it-yourself lighting

colorful metal dome shaped
12"- 20"D

akari Japanese paper globe
12"- 50"D

Paint a colorful switchplate

decorate switchplate
with stickers,
self-adhesive
plastic strips,
tape, or Magic
Markers
or remove plate to paint

protect wall
with masking
tape

butterflies and other
colorful shapes
pinned to lampshade

colored yarn
wound around
shade;
glue yarn at
start and finish

lamp making kit:
socket, cork,
cord,
plug

decorate bottle

sand inside
for weight

bottle lamp

quick paper
pleated shade

Child's ball

To apply self-adhesive plastic, stick new covering on a little at a time, smoothing out air bubbles. Overlap at top and bottom. To apply fabric, put glue or adhesive spray over the shade, and inside top and bottom to stick down overlap.

Fabric can also be sewn onto a shade or an old wire frame.

A quick and pretty lampshade cover can be made by pleating white paper or wallpaper to cover right over an old shade. Cut paper twice the circumference and cut to height; then pleat and punch holes 1" below the top and thread a ribbon, cord, or knitting wool through the holes. Then place it over old shade, pull tie tight, and glue or tape the open ends together (see illustration).

To hang a light fixture from the ceiling when there is no outlet on the ceiling: attach it to the ceiling and then run a long cord on ceiling to wall and down to an outlet. To avoid a tangle of cord underfoot, attach with insulated staples along the floor woodwork and up the wall next to moldings or shelves, where it will be less noticeable. Cord color should blend with walls and ceiling.

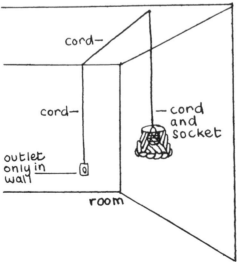

cord

cord

outlet
only in
wall

cord
and
socket

room

trim and hang a shade:
large tin can,
wicker basket,
cardboard
bucket

Floors

A free-flowing solid expanse of floor—one uninterrupted by area rugs or a change in floor covering—will go a long way toward making a room look larger. To add to this illusion, keep the floor neutral and low-key and avoid strong colors and patterns that will "advance" and make a space seem smaller; and "extend" the floor by painting baseboards, or even walls, the same color.

Now to the unending debate among parents as to which is preferable, a carpeted or a hard-surface floor. For cleanliness, solid or composition vinyl or a finished wood floor is easier to maintain, especially at a child's "messy stage"—and it's a firmer surface for running trucks, block-building, and crayoning. However, a carpeted floor is clearly better for crawling babies as long as they are not allergic to dust. It looks and feels warm, and it is soft and sound-absorbent. A low-pile, medium-tone carpet with some slight variation in color will show less dirt and will require less cleaning than very dark or light colors such as pink, blue, and yellow.

To compromise:

Use a very-low-pile synthetic "industrial" carpet—a good cross between the two.

Use both—half and half: carpet the sleeping area, not the sloppy (play) area. Use matching *or* contrasting colors. An

throw rugs made
from bathroom carpeting
or carpet remnants

with marker draw
in center of flower

glue on trim

cat: glue on felt
for eyes and yarn
for whiskers

alternative is to put down a piece of sheet flooring under the messy place, i.e., an easel or play table.

Use one or more colorful machine-washable inexpensive cotton throw rugs (so he doesn't trip on them, these should be removed when a baby starts to walk) over a hard surface floor, or one larger area rug that can be removed for cleaning.

Some parents say bare floors are best even though newly walking babies wearing socks or footed pajamas may slip and slide on them at first. If you already have a nice wood floor that is smooth and free of splinters, the most economical and practical approach may be to finish it with a coat of polyurethane (semigloss is less slippery than high gloss) or stain (colored or wood tone), or wax; or to cover it with several coats of deck paint. (Paint can chip and scratch, but worn spots are easy to touch up.)

How to Paint the Floor

Here, as in all other floor-furnishing projects, make sure the floor is clean (no dirt or grease), dry and smooth (little holes can be filled in with wood filler). Use a roller or large brush on a long stick to apply floor or deck enamel. After filling in perimeter with a small brush, start in the farthest corner and work toward the door, painting yourself out of the room. Repeat when dry, if necessary.

For extra protection of painted designs, cover the finished floor with a coat of polyurethane.

Special Painted Effects

Splatter-painting Shaking a brush full of another color paint over the painted floor is fun, and the finished floor will be less likely to show chipping.

A painted rug This requires no washing! Make one by simply marking off the outside perimeter of the desired shape with masking tape and then painting it in.

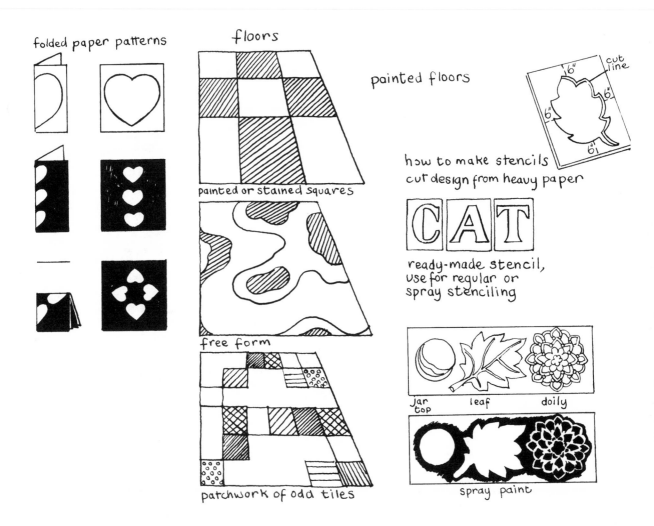

folded paper patterns

floors

painted floors

painted or stained squares

free form

patchwork of odd tiles

how to make stencils
cut design from heavy paper

CAT

ready-made stencil,
use for regular or
spray stenciling

jar
top   leaf   doily

spray paint

Stenciling   A wood floor regardless of condition can also be stenciled, and with it furniture, walls, fabric, the area along doors, windows, ceiling, and chair rails (see illustration). A stenciled floor is charming in a child's room and if it is properly sealed it will hold up very well. In fact, a stenciled design, like an antique, improves in appearance with wear. The design can cover the whole floor, only the center, or border the perimeter. Use Japan or acrylic paint or paint in a spray can (easier and sloppier). Buy a special stenciling brush, or make your own by cutting short any stiff-bristled brush. Stencils

stenciled room

come in contemporary as well as classic patterns: Early American, Art Nouveau, Art Deco, and Victorian. They are available ready-made at art supply stores, or you can cut your own by copying the desired motif, such as one in the fabric of the room, onto thin plastic or heavy paper.

How to Tile Floors

One of the advantages of tile over sheet material (resilient or carpet) is that it is reasonably easy to install yourself, especially over a small area. If the tile is 12" square, the square feet of floor space (length times width) tells you the number of tiles needed. The chart below shows how to estimate the amount of material needed if tiles are 9" × 9". Add on 5 to 10 percent for waste. Leftover resilient tiles will make hardy, washable, matching surfaces for chests, counters, and tables.

Sq. Ft. to Cover	10	11	12	13	14	15	16	17	18	19	20	21	22	23	24
No. of Squares	18	20	21	23	25	27	28	30	32	34	36	37	38	41	43
Sq. Ft. to Cover	25	26	27	28	29	30	31	32	33	34	35	36	37	38	39
No. of Squares	44	46	48	50	52	53	55	57	59	60	62	64	66	68	69
Sq. Ft. to Cover	40	41	42	43	44	45	46	47	48	49	50	51	52	53	54
No. of Squares	71	73	75	76	78	80	82	84	85	87	89	91	92	94	96
Sq. Ft. to Cover	55	56	57	58	59	60	61	62	63	64	65	66	67	68	69
No. of Squares	98	100	101	103	105	107	108	110	112	114	116	117	119	121	123
Sq. Ft. to Cover	70	71	72	73	74	75	76	77	78	79	80	81	82	83	84
No. of Squares	124	126	128	130	132	133	135	137	139	140	142	144	146	148	149
Sq. Ft. to Cover	85	86	87	88	89	90	91	92	93	94	95	96	97	98	99
No. of Squares	151	153	155	156	158	160	162	164	165	167	169	171	172	174	176

How to Wallpaper or ConTact-Paper the Floor

"Wallpaper" a nursery floor that receives little foot traffic. Measure the area to be covered and then cut pieces of wallpaper to size. Apply spray adhesive or brush white glue onto paper or floor. Then, starting from the farthest corner and working toward the front, smooth paper onto floor. Trim off excess at edges with a single-edge razor. For more protection, cover with one or two coats of polyurethane.

Homemade Rugs

Rejuvenate an old cotton rug with dye and/or sew or glue on new fringe. Make a new rug by sewing together rug samples, or by painting a heavy piece of canvas with acrylics.

Walls and Ceilings

I want walls that my child can write on with chocolate.

While some parents say walls should be a bland color with pattern in fabric and accessories, others insist they should be exciting and visually enriching. But all agree they must be washable! Nice walls can dress up the dreariest room and define the most limited areas—an alcove, one corner, a single wall. For a unified appearance that will visually enlarge the nursery, paint or paper walls with color or pattern (a small-scale pattern will make walls seem to recede) that ties in or repeats other elements in the room, such as curtains, carpet, and furniture. Or cover walls with matching fabric attached with diluted white glue (half glue, half water) or fabric adhesive; or hang by shirring on rods or by stretching the fabric and stapling or nailing it to the walls. Even if only a portion of the wall area is decorated, the effect can be total. So why not paint just the doors and moldings, or wallpaper just the ceiling or one wall?

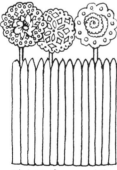

picket fence with green tape stems, and paper doilies for flowers.

Painted Walls

Dollar for dollar, paint is the most economical and effective wall treatment. Along with washable wall coverings, a painted wall is the easiest to maintain, and it is inexpensive to change—unless the color to be covered is very dark. There are two kinds of paint for the purpose. Latex, or water-based, paints are fast-drying, easy to spread, clean up quickly with soap and water, and are almost odorless. Alkyd, or oil-based, paints are thinned with a solvent (turpentine or mineral spirits), making them more resistant to water—a mixed blessing for many do-it-yourselfers who consider alkyd paint to be harder and messier to handle. Both are available in three finishes: flat, high gloss, and semigloss (also known as eggshell or satin finish). As its name implies, semigloss is a cross between flat and high gloss, combining the attributes of both and making it the best choice for nursery walls and woodwork under most conditions. The shinier a paint, the more washable, but more imperfections in a wall will show. (Imperfections show most under a high gloss, least under a flat paint. But flat alkyd is not washable at all and flat latex is only slightly better.) A high-gloss finish is crisp and contemporary, a little bit more expensive, and very durable—so it is also a good cover and camouflage for furniture. Since it reflects light, rooms painted with a high gloss paint will "glow" and seem to be more spacious. If you like shiny walls, use a high-gloss paint but use a flat paint on the ceiling.

Painting Tips

To calculate the amount of paint you will need, first calculate the number of square feet of wall to be covered in the entire room by multiplying the perimeter (the number of linear feet going all around the room) by the height of the room in feet. (The ceiling area will be the width times the length). To calculate the number of square feet of one wall, multiply the width by the height of that wall. The average gallon of paint covers roughly 350 to 400 square feet (covering power is usually on the label). Increase this amount if the ceiling, closets, built-ins, furniture, radiators, etc., are to be painted ; or if more than one coat will be necessary; or if the paint is thick (i.e., a textured

doing things with doors

dutch doors

hooks in hall outside room

letters cut from self-adhesive plastic

slate

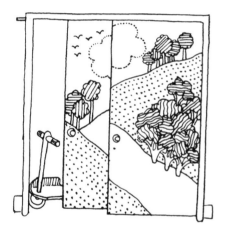

large mural on sliding door

paint) or the wall is very porous. Buy a good-quality paint, enough so that you can have leftovers for touch-ups, which you will surely have to do.

Paint in this order: ceiling and walls, doors, windows, trim (unless it is a different color, then paint first), floor.

Paint on the wall will look different than it does in the can (usually brighter) and in daylight or artificial light. To see how a color will look on a wall, make your own large sample by painting a large white card and taping it up to the wall (samples taken from paint stores are too small to tell much). Or buy a small amount of the paint—one or more quarts in different colors—and actually paint a small test portion of the wall (use a hairdryer to dry it fast and use the extra quart for accessories).

Since it can be custom-mixed, it is easier to match paint to carpet, wall covering, or fabric than vice versa, so select these first if possible.

You can paint over old wallpaper, even paper with a texture such as grass cloth, as long as it is firmly affixed to the wall—no peeling or buckling. But once having done this, you won't be able to remove it, which you probably would not want to do anyway.

Easy Ideas with Paint

Murals   Paint simple shapes, such as a geometric design; alphabets and numbers; squares and circles; clouds, rainbows and butterflies; animals, trees and flowers.

Paint lighter colors first so errors can be covered with subsequent darker colors. Get ideas from fabrics, children's books. Use ordinary household items for tracing, such as the bottoms of cups and saucers to make circles. (For the timid, two alternatives are: store-bought murals or murals you make from a material such as self-adhesive wall covering that can be quickly removed in the event of error.) And for those afraid to attempt a freehand mural, a mural can also be planned out on graph paper; just sketch or trace your design on it, then lightly sketch the grid on the wall with charcoal or chalk, then copy the design square by square. Keep the decoration simple— and low enough to be within a child's eye range.

Architectural detail   By painting doors, doorknobs, moldings, panels, even the ceiling a different or a contrasting color, you can add architectural interest to the room.

Textural surfaces   These hide dirt and are made with sponges, old toothbrushes, combs, or wads of paper run through wet paint or with textured stuccolike paint.

Stenciling   See preceding chapter, "Floors."

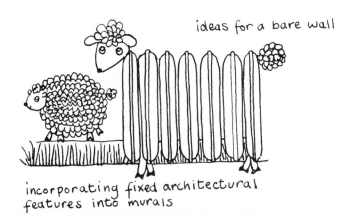

ideas for a bare wall

incorporating fixed architectural features into murals

homemade wall paper border, of cut-out dolls, and matching stencil shade.

Fabric-Covered Walls

On the surface, fabric does not seem to fill the bill as a practical wall covering for children's rooms; however, there are several advantages to using it, especially for an infant.

The same fabric used elsewhere in the room—as curtains or crib skirt, can be put on walls, producing a professional decorating appearance.

Fabric walls make a space feel cozy and warm.

Fabric covers imperfections and architectural features, such as cracks and moldings that cannot be removed.

Fairly thin, tightly woven cotton (or cotton and a synthetic mix) such as sheets or dress fabric, can be very inexpensive and resistant to soil—and perfect for matching window shades.

Fabric is relatively easy to put up and take down. It can be quickly pulled off walls (especially if it is stapled to a Sheetrock wall) and replaced with a more suitable washable, hard-wearing surface. Then the fabric can be used elsewhere.

If you are doing it yourself, and your workmanship is unpredictable, pick a fabric with a small, busy pattern. It will show fewer mistakes. To hide seams and staples, glue on decorative trim or nail on thin wood molding, painted to match or contrast.

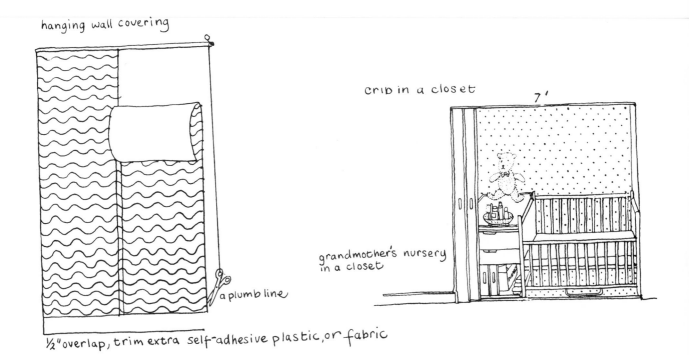

hanging wall covering

crib in a closet

7'

grandmother's nursery in a closet

½" overlap, trim extra self-adhesive plastic, or fabric   a plumb line

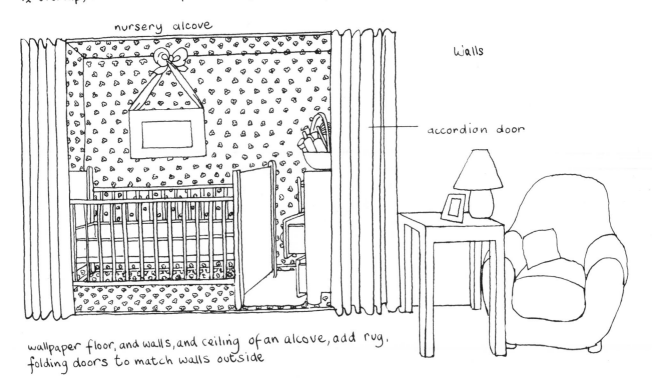

nursery alcove

Walls

accordion door

wallpaper floor, and walls, and ceiling of an alcove, add rug.
folding doors to match walls outside

Fabric can be put up several ways, and one or more methods may be used together:

Stretched and stapled, with a high-compression staple gun (staples will go into Sheetrock, wood, and some plaster walls), with tacks (harder to do), with double-faced tape (not as secure and the tape is expensive).

Pasted like wallpaper with white glue brushed first on nonporous walls (not on fabric).

Upholstered. The elegant way to "upholster walls" is to attach thin strips of wood (lattice) to the wall, put padding beneath, and tack fabric over onto the strips. The easier way is to skip the strips and staple the padding directly to the walls.

Shirred. Make "curtain walls" by sewing channels in top and bottom of fabric panel, then stretching and shirring on cord or rods that are attached to the top and bottom of wall.

How to Estimate the Amount of Fabric Needed

Again, be sure to have enough. Leftovers can always be used to trim accessories. To estimate how much: divide the perimeter of the room (or length of the walls if you are not doing a whole room) by the width of the fabric, subtracting about 6" for overlap. This will give you the number of widths or panels needed (for fabric that will be shirred, multiply by 2 or 3). The length of each panel will be the height of the area to be covered (usually the height of the room) plus about 3" to 4" for turning under at top and bottom. For example, multiply 10 widths by the length of each 9 feet (3 yards) and you will need 30 yards. Allow for more fabric if it has a large pattern that must be matched and less if there are several openings (windows, doors, fireplaces) in the room.

Wallpapered Walls

The word "wall*paper*" is a misnomer. There is a very wide variety of materials to cover walls and many are not paper at all. "Wall *covering*" is a more accurate description. For a child's room, a washable vinyl that can be all cloth or paperbacked is your best bet. Self-adhesive plastic (you may know it as Con-Tact paper) is another good choice, particularly if you are doing it yourself (although it is possible to become hopelessly entwined with a sticky panel that adheres to you instead of the wall). Covering only one wall with ConTact may lessen the frustration, and is effective if some of the furnishings in the room are also covered with it. Another advantage to self-adhesive plastic is that, unlike paper, it "gives," permitting surfaces not usually suitable for papering, such as old furniture or bathroom tile, to be covered. Coverings are available scrubbable, that is, with a washable coating; strippable, easily removed when you want a change; and prepasted and/or pretrimmed for easy installation. Strippable wallpaper borders 8" to 10" wide are popular for nurseries now because they are very decorative, easy to put up, take down, and replace with a more mature motif.

"Papering" cannot cover a very bad wall and can be an expensive proposition. Prices vary enormously, especially if you are adding in the cost of labor, which you should in the absence of abundant time, talent, and patience. *Note:* One way to cut down on the work and the cost is to cover only part of the room—say, the ceiling or only the upper half of the wall. Nail on some molding to make a chair rail, then paint below and paper above. This has the added advantage of keeping the paper away from the part of the wall that gets the most wear and tear.

How to Estimate Wallpaper Needs

Wallpaper is made in single, double, or triple rolls. To estimate the number of rolls, do the following or refer to the chart on page 136. Each single roll of wallpaper has 36 square feet of material. About 30 square feet of it is usable, allowing for waste and matching of patterns. Divide the number of square feet to be covered by 30 to find out how many rolls you need. Then deduct one single roll for every two openings, or two single rolls for every three openings: windows, doors, and fireplaces. Leftover wallpaper can be used for patching and to cover or trim furniture, such as a desk top (vinyl coated "paper" is best) and accessories such as lamps and picture frames.

Ceiling Height	8 Feet	9 feet	10 feet	11 Feet	12 Feet	Single Rolls
Size of Room Feet	SINGLE ROLLS					Needed for Ceiling
8 × 10	9	10	11	12	13	3
10 × 10	10	11	13	14	15	4
10 × 12	11	12	14	15	16	4
10 × 14	12	14	15	16	18	5
12 × 12	12	14	15	16	18	5
12 × 14	13	15	16	18	19	6
12 × 16	14	16	17	19	21	6
12 × 18	15	17	19	20	22	7
12 × 20	16	18	20	22	24	8
14 × 14	14	16	17	19	21	7
14 × 16	15	17	19	20	22	7
14 × 18	16	18	20	22	24	8
14 × 20	17	19	21	23	25	9
14 × 22	18	20	22	24	27	10
16 × 16	16	18	20	22	24	8
16 × 18	17	19	21	23	25	9
16 × 20	18	20	22	24	27	10
16 × 22	19	21	23	26	28	11
16 × 24	20	22	25	27	30	12
18 × 18	18	20	22	24	27	11
18 × 20	19	21	23	26	28	12
18 × 22	20	22	25	27	30	12
18 × 24	21	23	26	28	31	14

Homemade Wall Covering and Other Interesting Effects

Cover one or more walls with the following (use diluted white glue and protect with polyurethane):

Comic pages, family photographs, magazine covers.

Wrapping or brown paper.

Travel or art posters. Put them inside moldings, if you have them, and they will look "framed."

Cutouts from wallpaper or fabric can be used to decorate walls and everything else in sight.

Simulate or accent architectural features by gluing on felt, trim, ribbon, tape, blanket binding, pieces of fabric, almost anything, in strips or narrow rolls, to make chair rails, picture moldings, and use around walls, doors, and windows. (Add the same trim to a bedspread or curtain or shade.)

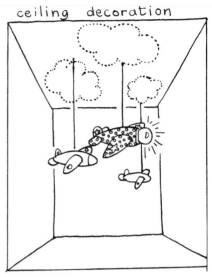

ceiling decoration

Small planes are decoration, center one is a light fixture

Ceilings

A baby's view of the world is largely from his back looking up, so he will surely enjoy decorations above the crib. Mobiles are not the only hanging decorations. Lightweight stuffed or inflated dolls, animals, balloons, kites, or flags can dangle from strings attached to the ceiling. And why stop at just covering the crib with a canopy when the whole ceiling can be draped or tented with fabric? More simply, the ceiling can be painted in a contrasting color for a more decorative wall as well as practical effect: lighter colors will make the ceiling seem to recede and thus appear higher, while the darker shades will make it "advance" and seem lower.

Easy Ideas for Ceilings

Ceiling moldings or beams can be brought out with paint in a contrasting color.

The ceiling can be decorated with paint or self-adhesive

plastic cutouts in an appropriate motif: stars, clouds, birds, butterflies.

The ceiling can be fully "wallpapered," or just a wallpaper frieze or border can be added where ceiling and wall meet.

Do-It-Yourself vs Hiring Help

Not everyone is cut out to be a painter or a carpenter, even in this era of the spray can and staple gun. These words of wisdom—and warning—from David Stiles, a well-known industrial designer, on making children's furniture seem to apply to all decorating projects:

> *To Buy or to Build—that is the BIG question (even for me, an industrial designer with a shop full of expensive tools and a child that keeps on growing). You simply have to weigh the differences by asking yourself: Do you have the time it takes to build something? (It generally takes twice as long as you first figure.) Are there special tools that you need or special skills that you don't have?*
>
> *What about all the mess and sawdust?*
>
> *But—Ahh, the Advantages!*
>
> *You can build something that fills the space perfectly—no disturbing gaps that you get when you buy store-bought furniture. You can build something UNIQUE that is impossible to find elsewhere that is adapted to your very own child. You can build it so that it is adjustable and grows with the child, or you can build it so that it can be changed at a later date to fill a different function. And you can build it NOW, without the customary wait of four to six weeks delivery. (Even then you might be putting it together yourself—if it's*

knocked down.) You can build it a lot cheaper than it costs and you can use the leftover scraps for other projects. But the most convincing point of all is to hear your child say "You wanna see what my Mom and Dad made for me?"

Working with Workmen

If you choose or find it necessary to obtain assistance, here are some tips on hiring and handling help.

The best way to find a workman—painter, carpenter, even an interior designer—is through word-of-mouth. In addition to answers about the quality of his performance, you can find out about other intangibles such as whether he is reliable, messy, pleasant, cooperative, imaginative, and—most important—accurate in making estimates. If you have heard about or seen someone's work but do not have a personal contact, ask for references. I also respond favorably to professionals who act it—that is, they return your call promptly, arrive on time for appointments, and so forth. And I recommend comparison shopping and hiring experienced amateurs such as students for routine interior work.

Once you have found a workman, determine the cost and the method of payment. All estimates should be written down and should specify the complete cost, including materials such as wood or paint, which are often extra, when (day and *time*) the job will be started and—most important to expectant parents—finished (withhold that final payment until it is). You will also need to be sure that the working hours correspond to your routine or your building's rules.

Changes are expensive and cause delays, so try to plan ahead. Once the job has started, try to be there to supervise.

Selecting an Interior Designer

Whatever the motive, it may be that you have decided to hire an interior designer and—depending upon your particular situation—it may be well worth it: "Roger's parents knew that he felt he was second in importance to his older sister and they wanted him to have a room that he could feel autonomous and special in. It worked. At the completion of his room, he invited his entire kindergarten class to visit his new space." So says Lila Schneider, designer.

To the tips for hiring others, I would add the following:

Compatibility of personality and taste is crucial. So you should both meet the designer and see his/her work, even if only in pictures.

Establish the cost (sometimes an hourly fee, sometimes a percentage of purchases or the cost of construction) and method of payment. Unless he/she is very famous, you do not necessarily have to pay more for a talented designer than an unimaginative one.

Depending upon his or her schedule, a designer may be able to spring into action and relieve you of the burden of measuring, planning, shopping, ordering, and finally, supervising both the installation and the other help.

Another advantage in hiring a designer is that he or she usually knows the market and where to buy everything. When it comes to babies' rooms, I am not so sure this is true. Since relatively few designers decorate nurseries (for the reasons already mentioned) or even have children, you will probably know more than most of them do. On the other hand, it is important to remember that you are actually designing more than one room. A good professional can give you valuable ideas about the configuration of your whole house and how it can be adapted to and expanded for your new needs. You may not need full-time help for this but rather an hourly-basis consultation with an expert. In some cities this service is available for a modest flat fee from independent designers or from the decorating department of larger stores.

Conclusion

Whether you do it by yourself or call on professional help, designing and decorating your baby's space can enhance the joys of parenthood, and if done well, the pleasures of childhood. A nice space is an expression of affection and caring. You and your child deserve no less, and with planning and know-how you can create the right environment for your child, however limited your space and budget. Good luck!

Baby Space Planner

Shopping Sources (Stores, Internet Sites, Retail Catalogs)

For Furniture and Equipment:

STORES

There are many *baby specialty stores,* including large ones like Baby Superstore in Stamford, Connecticut; the small but excellent chain Buy Buy Baby in Scarsdale, New York, Rockville, Maryland, and Paramus, New Jersey, and in New York City, FAO Schwartz, Wicker Wonderland, Albees, and Ben's.

Shopping sources also include *department stores, regular furniture stores, chains* such as IKEA, LL Bean, Home Depot, Pottery Barn, and *discount chains* with baby departments such as Walmart, Target, Kmart, and baby chains such as Baby's R Us (888-BABYSRUS) or Baby Depot (800-444-COAT). Many stores now have Internet sites and catalogs.

INTERNET SITES AND CATALOGS

You can now shop conveniently by phone, fax, or computer—perhaps after visiting a store—at any time, all day, every day, when you are housebound baby-sitting or up in the middle of the night nursing. (Surfing can be slow, so it is always efficient to be doing something else at the same time!)

Below are some listings to start off, but keep in mind that catalogs and sites change frequently. Look for additional sites using your search engine and key words such as *baby furniture, equipment, floor covering, curtains,* etc. Keep in mind that these sources may be more suitable for ordering equipment and accessories than larger, bulky furniture such as cribs and chests, for which shipping costs may be high and there is a chance of damage and even more delay than usual.

Here are some tips for cybershopping (and most of these apply to catalog shopping as well): check on delivery and return policies, including delivery, handling and returns, how long delivery takes, and whether purchases can be shipped quickly for additional cost. Confirm delivery date so that *your* delivery doesn't happen before the bassinet arrives and ask to be e-mailed if a desired item is out of stock. Be sure to add handling, delivery, and tax to the cost. Inquire about discounts, sales items, gift wrapping, gift certificates, etc.

For greater protection later on, if necessary, use your credit card, but check that the site offers secure ordering using encryption technology. It is always a good idea to print out your order confirmation, to keep any other records of purchase, and to get a toll-free number connecting to a live person should you need to track your order.

Safety & Product Evaluation Services

Child Secure
10660 Pine Haven
N. Bethesda, MD 20852
1-800-450-6530

National Safety Council
1121 Spring Lake Drive
Itasca, IL 60143
1-800-621-7619

U.S. Consumer Product Safety Commission
Washington, DC 20207
1-800-638-2772 (for information on latest recalls)
www.cpsc.gov

Juvenile Products Manufacturers Association
236 Route 38 West, Suite 100
Moorestown, NJ 05057
(609) 231-8500 (Call for a list of certified products or write for their brochure *Safe and Sound for Baby*. Include a stamped, self-addressed envelope.)
www.JPMA.org

National Parenting Center
22801 Ventura Boulevard, Suite 110
Woodland Hills, CA 91367
1-800-753-6667
www.TNPC.com

The Danny Foundation (crib safety information)
www.dannyfoundation.com

Consumer Reports
www.consumerreports.org

Better Business Bureau
www.bbb.org

National Periodicals

American Baby
249 West 17th Street
New York, NY 10011
www.americanbaby.com

Baby Talk
1325 6th Avenue
New York, NY 10019
www.parenting.com

Child
375 Lexington Avenue
New York, NY 10017
(212) 449-2000

Mothering
P.O. Box 1690
Santa Fe, NM 87504
1-800-984-8116
www.mothering.com

Parenting magazine
301 Howard Street
San Francisco, CA 94105

Parents magazine
375 Lexington Avenue
New York, NY 10017
1-800-727-3682

Fit Pregnancy
P.O. Box 37266
Boone, IA 50037
www.fitpregnancy.com

Retail Catalogs

To find and order other catalogs on the web:
www.catalogsite.com

Baby Catalog of America
738 Washington Avenue
West Haven, CT 06516
1-800-PLAYPEN
www.babycatalog.com

Biobottoms
100 Plaza Drive
Seacaucus, NJ 07094
1-800-766-1254
www.biobottoms.com

Eddie Bauer (bedding)
P.O. Box 182639
Columbus, OH 43218
1-800-426-8020
www.eddiebauer.com

Garnet Hill (bedding)
231 Main Street
Franconia, NH 03580
1-800-622-6216

JC Penny (Pennies from Heaven catalog)
1-800-222-6161
www.JCPenney.com

Jeannie's Kids Club
7245 Whipple Avenue NW
North Canton, OH 44720
1-800-363-0500
F (216) 494-0265
www.kidstuff.com

Just Pretend
104 Challenger Drive
Portland, TN 37148-1729
1-800-286-7166
www.justpretend.com

The Land of Nod (bedding and accessories)
P.O. Box 1404
Wheeling, IL 60090
1-800-933-9904

Land's End (bedding)
1-800-345-3696
www.landsend.com

One Step Ahead
75 Albrecht Drive
Lake Bluff, IL 60044
1-800-274-8440
F (847) 615-7236
www.onestepahead.com

Perfectly Safe
7835 Freedom Avenue NW #3
North Canton, OH 44720
1-800-837-5437
www.kidsstuff.com

Pottery Barn Kids
7720 North West 86th Street
Oklahoma City, OK 73132
(405) 717-6000

The Right Start Catalog
1-800-548-8531
F 1-800-762-5501
www.rightstart.com

Schweitzer Linen
1053 Lexington Avenue
New York, NY 10021
1-800-554-6367
www.schweitzer-linen.com

Internet Sites

(See also Retail Catalogs and Safety & Product Evaluation Services for additional sites)

BABY FURNITURE IN GENERAL

www.ababy.com

www.abcparenting.com

www.babiesrus.com (1-800-BABYRUS)

www.babybestbuy.com

www.babybusiness.com

www.babycenter.com (1-877-551-BABY)

www.babyfurniture.com

www.babygear.com

www.babylulu.com

www.babyonline.com

www.babyresource.com

www.babystyle.com (1-877-378-9537)

www.babyuniverse.com

www.babyzone.com

www.bestbabyproducts.com

www.childrensfurniture.com

www.cyberbabymall.com

www.geniusbabies.com

www.ibaby.com

www.maternitymall.com

www.maternityzone.com

www.mommy-mall.com

www.parenthoodweb.com

www.parents.com

www.parenttime.com

www.royalbaby.com

www.rightstart.com

www.thebabynet.com

www.20ishparents.com

SPECIFIC FURNITURE, FURNISHINGS, EQUIPMENT & TOYS

www.babybargainsbook.com

www.babybox.com (for gifts)

www.bcfdirect.com (1-800-144-2678)

www.cradlesofdistinction.com (1-800-961-0991)

www.etoys.com (1-800-463-8697)

www.fisherprice.com

www.faoschwartz.com (1-800-791-1141)

www.gymboree.com

www.kbkids.com (1-877-452-5437)

www.kidstuff.com

www.stephanieanne.com (furniture, 1-888-885-6700)

www.toysrus.com (1-888-869-7932)

GENERAL FURNITURE (WITH BABY DEPARTMENTS)

www.babydepot.com

www.decoratetoday.com

www.EthanAllen.com

www.furniture.com (offers a comprehensive list of manufacturers with web sites)

www.goodhome.com

www.ivillage.com

www.homepoint.com

www.living.com

www.marthastewart.com

www.women.com

NOTE: Some sites now offer online experts to help. Sites, sometimes called "bots," that scour the web to find the best prices include:

www.dealtime.com

www.mysimon.com

www.pricescan.com

Nursery Furniture Templates

Scale: $\frac{1}{2}$″ = 1′
See: How to Arrange Juvenile Furniture and Equipment, *page* 21
Cut-out templates (cut smaller, as necessary)

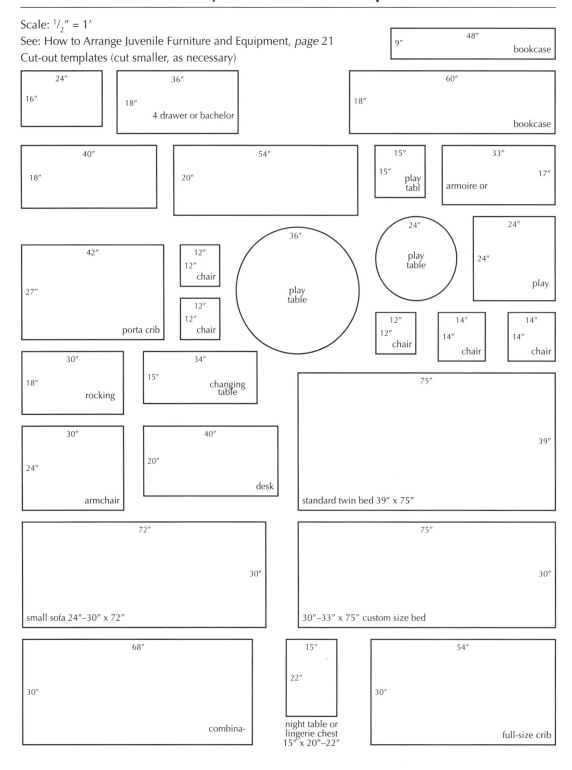

9″ — 48″ — bookcase

24″ / 16″

36″ / 18″ / 4 drawer or bachelor

60″ / 18″ / bookcase

40″ / 18″

54″ / 20″

15″ / 15″ / play tabl

33″ / 17″ / armoire or

42″ / 27″ / porta crib

12″ / 12″ / chair

12″ / 12″ / chair

36″ / play table

24″ / play table

24″ / 24″ / play

30″ / 18″ / rocking

34″ / 15″ / changing table

12″ / 12″ / chair

14″ / 14″ / chair

14″ / 14″ / chair

75″ / 39″ / standard twin bed 39″ x 75″

30″ / 24″ / armchair

40″ / 20″ / desk

72″ / 30″ / small sofa 24″–30″ x 72″

75″ / 30″ / 30″–33″ x 75″ custom size bed

68″ / 30″ / combina-

15″ / 22″ / night table or lingerie chest 15″ x 20″–22″

54″ / 30″ / full-size crib

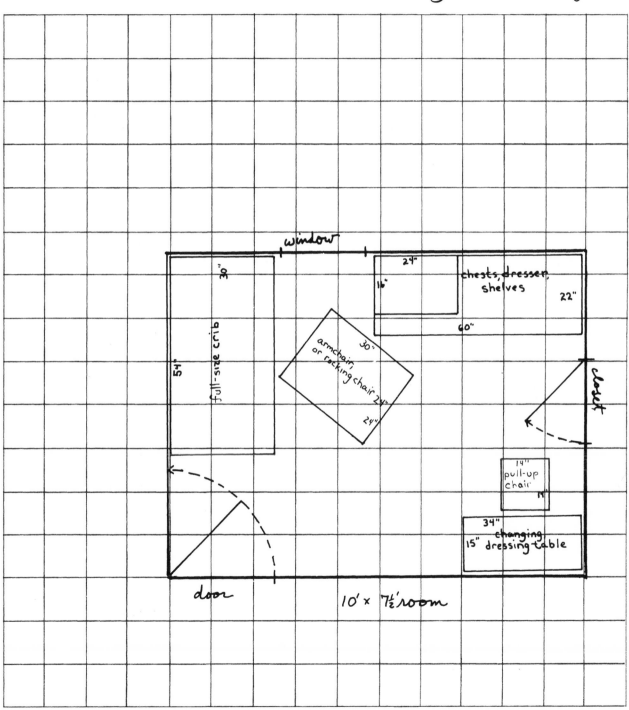

window

30"

54"

full-size crib

24"

16"

chests, dresser, shelves

22"

60"

armchair, or rocking chair 24"

30"

24"

closet

14"
pull-up
chair

14"

34"

changing
15" dressing table

door

10' x 7½' room

Notes, Measurements, Sketches

Notes, Measurements, Sketches